D1272514

New Media, 1740–1915

Media in Transition

David Thorburn, series editor

Edward Barrett, Henry Jenkins, associate editors

New Media, 1740–1915, edited by Lisa Gitelman and Geoffrey B. Pingree, 2003

Democracy and New Media, edited by Henry Jenkins and David Thorburn, 2003

Rethinking Media Change: The Aesthetics of Transition, edited by David Thorburn and Henry Jenkins, 2003

New Media, 1740–1915

Edited by Lisa Gitelman and Geoffrey B. Pingree

The MIT Press
Cambridge, Massachusetts
London, England

This book was set in Perpetua by Graphic Composition, Inc.

Printed and bound in the United States of America.

Library of Congress Cataloging-in-Publication Data
New media, 1740–1915 / edited by Lisa Gitelman and Geoffrey B. Pingree.
     p. cm. — (Media in transition)
   Includes bibliographical references and index.
   ISBN 0-262-07245-9 (hc. : alk. paper)
    1. Mass media—History. I. Gitelman, Lisa. II. Pingree, Geoffrey B. III. Series.

  P90 .N5 2003
  302.23'09—dc21

                                        2002029542

Contents

Series Foreword

David Thorburn, editor
Edward Barrett, Henry Jenkins, associate editors

New media technologies and new linkages and alliances across older media are generating profound changes in our political, social, and aesthetic experience. But the media systems of our own era are unique neither in their instability nor in their complex, ongoing transformations. The Media in Transition series will explore older periods of media change as well as our own digital age. The series hopes to nourish a pragmatic, historically informed discourse that maps a middle ground between the extremes of euphoria and panic that define so much current discussion about emerging media—a discourse that recognizes the place of economic, political, legal, social, and cultural institutions in mediating and partly shaping technological change.

Although it will be open to many theories and methods, three principles will define the series:

- It will be historical—grounded in an awareness of the past, of continuities and discontinuities among contemporary media and their ancestors.
- It will be comparative—open especially to studies that juxtapose older and contemporary media, or that examine continuities across different media and historical eras, or that compare the media systems of different societies.
- It will be accessible—suspicious of specialized terminologies, a forum for humanists and social scientists who wish to speak not only across academic disciplines but also to policymakers, to media and corporate practitioners, and to their fellow citizens.

New Media, 1740–1915
edited by Lisa Gitelman and Geoffrey B. Pingree

Proceeding from the truism that all media were once new media, the essays in this collection explore media as they emerged in the eighteenth, nineteenth, and early twentieth centuries. The media addressed range from familiar examples, such as telephones and phonographs, to unfamiliar curiosities, such as the physiognotrace and the zograscope.

Exploring moments of transition when each new medium was not yet fully defined, its significance in flux, these essays aim to clarify our understanding of the specific material and historical environments in which new media emerge and of the ways in which habits and structures of communication are naturalized or normalized.

Acknowledgments

New Media, 1740–1915 is a genuinely collaborative project, one that arose from enthusiasms widely shared. We wish to acknowledge those institutions and thank those individuals who have in some way helped us build it. We are grateful to David Thorburn and Henry Jenkins at MIT for first taking an interest in the book and for supporting a project that includes a range of young and relatively unknown scholars. Our contributors are indeed collaborators, and we thank them for their interest and dedication.

For material and financial support throughout the development and execution of this work, we would like to thank Catholic University—especially Antanas Suziedelis, dean of the School of Arts and Sciences, and the Office of Sponsored Research—as well as the Obermann Center for Advanced Studies at the University of Iowa. We would specifically like to thank our colleagues Spencer Cosmos, Glen Johnson, and Steve McKenna in the Program in Media Studies at Catholic University, with whom we have been engaged for several years in a conversation about media that has greatly enriched our thinking. Without them media studies, not just *New Media*, would be a lesser inquiry (and a lot less fun).

For their practical assistance, we would like to thank Bridget Brennan for her help in assembling the parts of this whole, Mearah Quinn-Brauner for proofreading, and Lisa Abend for her helpful insights and keen editing eye.

Finally, for moral support we would like to thank Lisa Abend, Judy Babbitts, Pat Crain, Greg Pingree, Laura Rigal, Gayle Wald, and the many friends and acquaintances we cajoled into reading our proposal along the way.

Introduction: What's New About New Media?

In the short space of a current college student's lifetime, the internet has gone from a specialized, futuristic system to the network that most significantly structures how we engage daily with the world at large. It is now obvious to anyone who uses a computer that intellectual exercises as basic as reading the newspaper or doing research have become fundamentally different activities largely because of the internet. So too have our views of communication in general; the very notion of globalization, so consuming in today's world, is predicated on the possibilities engendered by a technology barely twenty years old. Such is the nature of "new media." Computers, and the digital systems and products for which they are currently a shorthand, are what most of us think of when we hear the words *new media*. And why not? The world of computer hardware, software, email, and ebusiness is for most of us the latest communication and information frontier. Part of our experience of digital media is the experience of their novelty.

Yet if we were asked to think of other "new media," we might have a harder time coming up with obvious examples. We would have no problem citing instances of "old media": typewriters, vinyl record albums, eight-track magnetic tapes, and the like. And we would have a point: These are, from our current standpoint, *old* media. But they were not always old, and studying them in terms that allow us to understand what it meant for them to be new is a timely and culturally important task, an exercise that in this volume we hope profitably to apply to media much older than we are.

As our title suggests, this collection of essays challenges the notion that to study "new media" is to study *today's* new media. All media were once "new media," and our purpose in these essays is to consider such emergent media within their historical contexts—to seek out the past on its own passed terms. We do so, in part, to counter the narrow devotion to the present that is often evident today in "new media" studies, a growing field whose conceptual frameworks and methods of inquiry are heavily influenced by experiences of digital networks and the professional protocols of the social science of communications. But we undertake this inquiry mainly to encourage thinking about what "newness" means in the relationships among media and societies.

There is a moment, before the material means and the conceptual modes of new media have become fixed, when such media are not yet accepted as natural, when their own meanings are in flux. At such a moment, we might say that new media briefly acknowledge and question the mythic character and the ritualized conventions of existing media, while they are themselves defined within a perceptual and semiotic economy that they then help to transform. This collection of essays explores such moments in order to enrich our contemporary perspective on what media are, and on when and how they are meaningfully "new."

New Media, 1740–1915 focuses on the two centuries before commercial broadcasting because its purpose is, in part, to recuperate different (and past) senses of media in transition and thus to deepen our historical understanding of, and sharpen our critical dexterity toward, the experience of modern communication. Indeed, we have marked the years between 1740 and 1915 as boundaries for our project because this period is crucial to understanding how electronic and digital media have come to mean what and how they do. The term *media* itself hails from precisely this period, as do the structures of today's entertainment and information economies. Thus, the media forms and practices studied in this collection are "new" in a double sense: First, they newly receive the scholarly attention they deserve; and second, they are considered within their original historical contexts, their novelty years. In this, these essays provide a new perspective on the meaning of "newness" that attends to all emerging media, while they also tell us something about what all media have in common.

Yet our intention is not only to acknowledge the initial novelty of diverse media, but also to understand better how such media acquire particular meanings, powers, and characteristics. Drawing from Rick Altman's idea of "crisis historiography," we might say that new media, when they first emerge, pass through a phase of identity crisis, a crisis precipitated at least by the uncertain status of the given medium in relation to established, known media and their functions.[1] In other words, when new media emerge in a society, their place is at first ill defined, and their ultimate meanings or functions are shaped over time by that society's existing habits of media use (which, of course, derive from experience with other, established media), by shared desires for new uses, and by the slow process of adaptation between the two. The "crisis" of a new medium will be resolved when the perceptions of the medium, as well as its practical uses, are somehow adapted to existing categories of public understanding about what that medium does for whom and why.

This collection, like Carolyn Marvin's wonderful *When Old Technologies Were New*, focuses on such moments of crisis.[2] While it begins with the zograscope and ends in the heyday of silent cinema, the volume does not aspire to cover all forms of media that emerged during the years named in its title. Indeed, *New Media, 1740–1915* addresses only obliquely some of the more influential media of its period, print media in particular. Most of the following essays (unlike Carolyn Marvin's work) focus on media—zograscopes, optical telegraphs, the physiognotrace—that failed to survive for very long. They are, in Bruce Sterling's words, today's "dead media."[3] Yet because their "deaths," like those of all "dead" media, occurred in relation to those that "lived," even the most bizarre and the most short lived are profoundly intertextual, tangling during their existence with the dominant, discursive practices of representation that characterized the total cultural economy of their day.

Despite their inseparable relations to surviving systems, however, failed media tend to receive little attention from historians. "Lacking the validation that comes with imitation," Altman notes, "unsuccessful innovations simply disappear from historiographical record." His suggested corrective for this excessive focus on, for example, "cinema-as-it-is," is an attention to "cinema-as-it-could-have-been" or "cinema-as-it-once-was-for-a-short-time-but-ceased-to-be." *New Media* aims to apply some of this "could-have-been" and "was-for-a-short-time" kind of thinking to past new media. Because our understanding of what media are and why they matter derives largely from our understanding and use of the media that survived—those devices, social practices, and forms of representation with which we interact every day—the importance of this kind of analysis is easy to overlook.

By getting inside the "identity crises," by exploring the "failures" (in some cases) of older new media, the essays in this collection will help to counter what Paul Duguid has warned are two reductive "futurological tropes" characteristic of the experience of modern media. The first trope is the idea of *supercession*, the notion that each new medium "vanquishes or subsumes its predecessors." From this idea follows the current belief that in the digital age the book is doomed, or, according to the peculiar auguries of earlier times, the conviction that typewriters would replace pens or that radios would replace phonographs. The second futurological trope is the idea of increasing *transparency*, the assumption that each new medium actually mediates less, that it successfully "frees" information from the constraints of previously inadequate or "unnatural" media forms that represented reality less perfectly.[4] This notion—that because of their greater transparency, newer media supersede their predecessors—shapes both the experience and the study of media today. Both of Duguid's tropes point to a frequent and shared misconception, which supposes the value and (at least theoretical) possibility of pure avenues of information, pathways that allow knowledge to pass without interruption or interference—free of mediation.

This assumption creates an interesting paradox. The best media, it would seem, are the ones that mediate least. They are not, as we think of them, media at all. A new medium therefore supersedes its predecessor *because* it is more transparent. Few would disagree, for example, that a conversation with a friend on the telephone allows for a greater exchange of personal, idiosyncratic information than a dialogue conducted via telegraph. And to a large degree, this thinking is persuasive. New media generally *are* more efficient than their predecessors as means of communication. Yet there is more to understanding what happens when people communicate through a given medium than merely ascertaining what level of accuracy and amount of data the exchange involves. This observation—that there is more than accuracy and amount to any exchange—comprises a founding rationale for the field of media studies, whether characterized aphoristically by Marshall McLuhan ("the medium is the message") or more recently expressed (and complicated) in Derridian terms, that the *supplement*—the "specific characteristics of material media"—can never be "mere" supplement; it is "a necessary constituent of [any] representation."[5] To put it simply, looking for content apart from context just won't work.

Owing in part to the linear progress unthinkingly ascribed to modern technology, media (so often referred to portentously as *the* media) tend to erase their own historical contexts. Whether shadows in a darkened cave or pixelated images on a luminous monitor, the media before us tend, anachronistically, to *mediate* our understanding of their past. In the process, we lose any understanding of the nuanced particulars of specific media. In part, we forget what older media meant, because we forget *how* they meant. Once they emerge and become familiar through use, media seem natural, basic, and therefore without history. Of course we say "Hello?" when we answer the telephone; of course we hear a dial tone when we pick it up to place a call. Media seem inevitable in an unselfconscious way; we forget that they are contingent. Alexander Graham Bell apparently wanted people to say "Ahoy!" when they picked up the phone, but English speakers settled on "Hello?" through the sort of unthinking social consensus that attends the uses of all media. In a similar fashion, the dial tones, 12-volt lines, and modular jacks we use today all were shaped historically by a complex of forces—technological, to be sure, but also social, economic, and representational.

When we forget or ignore the histories of each of these new media we lose a kind of understanding more substantive than either the commercially interested definitions spun by today's media corporations or the causal plots of technological innovation offered by some historians.[6] For example, it is undoubtedly important to be able to note, as many scholars have, how the invention of the cinema is linked to past practices of, say, lecturing with slides, as well as how it predicts certain elements of future practices. But what we often overlook are the kinds of things that only a deep analysis of specific media cases can offer—how interpretive communities are built or destroyed, how normative epistemologies emerge. No medium new or old exists as a static form. Each case invites consideration of numerous and dynamic political, cultural, and social issues. We might say that, inasmuch as "media" are media of communication, the emergence of a new medium is always the occasion for the shaping of a new community or set of communities, a new equilibrium.

As we have suggested, when a new medium is introduced its meaning—its potential, its limitations, the publicly agreed upon sense of what it does, and for whom—has not yet been pinned down. And part of the lure of a new medium for any community is surely this uncertain status. Not yet fully defined, a new medium offers possibilities both positive (one of our authors argues that zograscopes helped construct polite society) and negative (another traces the threat telephones posed to Amish communities). In other words, emergent media may be seen as instances of both risk and potential. Today, for example, the internet offers unprecedented possibilities for global villages to coalesce, even while it threatens national or ethnic cultural traditions and provokes anguished discussions of privacy in a "connected" age. The same sorts of issues and anxieties surrounded the emergence of other media. Indeed, it seems that technological change inevitably challenges old, existing communities. The particulars of each case, however, are valuable to our larger understanding of how media help to shape and reshape culture.

Essays in this collection therefore examine media as socially realized structures of communication, where communication *is* culture—as James Carey explains it—a cultural process that involves not only the actual transmission of information, but also the ritualized collocation of senders and recipients.[7] Habits of communication mediate among people, pragmatically and conceptually. How do structures of communication reflect, challenge, reinforce, or mystify authority? How do they help imagine community? How do they help construct the aesthetic, or the mimetic? How do they orient the production and experience of meaning? How do they acquire and carry epistemological authority? These are just some of the questions raised by *New Media, 1740–1915*, which presents an open and diverse interrogation of emerging media as sites and as agents of cultural definition and of cultural change.

Ultimately, then, this is a book about framing: about how particular habits and media of communication frame our collective sense of time, place, and space; how they define our understanding of the public and the private; how they inform our apprehension of "the real"; and how they orient us in relation to competing forms of representation. We have selected the cases of new media that follow because they support these inquiries, casting such habits and media into relief, affording a vantage point from which better to see how cultural meanings are negotiated. But this collection is also about how we frame our own discussions of new media, for if this interrogation of emergent media is genuinely to illuminate our understanding of cultural definition and of cultural change, then we must be responsible about our own language. We must, in other words, acknowledge the key terms that are in play in our own discussions and attempt to define and deploy them as precisely as possible, not only for us now, but as they were used in earlier—and different—contexts.

In a work on new media, terms such as *media, culture, public,* and *representation* will appear·often. But insofar as this collection seeks to understand how the very idea of "media" evolves over time, we wish to employ such critical terms with care and to bring questions about their use and meaning squarely into the discussion itself as it proceeds. Our use of the word *technology* is a good example. This term denotes, as Leo Marx suggests, a necessary but "hazardous concept"; in this book the term helps organize our thinking about the material, instrumental conditions of modern life, yet for many readers it will also come larded with less considered shades of meaning, assumptions about "Progress" with a capital "P," or about technology as a preeminent cause in history.[8] Thus although we rely on this term as an organizing device in this collection (the essays proceed from technology to technology as a form of convenience), we also wish to urge particular awareness of its hazards. Likewise with other key critical terms. We know that we cannot exhaustively define "media," for instance, any more than we can completely pin down "culture" (a notion that is, as Naomi Mezey observes, "everywhere invoked and virtually nowhere explained"). Indeed, the cases we offer are about culture as struggle and media as means in that struggle—a fabric continually rewoven according to the interests of a given time and place. Rather than fixing such terms and pinning them to moving targets, however, we can frame our discussions of such pervasive concepts in self-conscious ways that make our attempts to understand them more useful.[9]

In this volume we offer cases that foreground the relationship between material and idea, between what people think or believe or wish and what they feel with their hands or see with their eyes or hear with their ears. Each of the essays in the collection thus reveals, in some fashion, the strong relationship between the contexts for some material, technological development, and shifts in self-imagining and public understanding. Erin Blake, Wendy Bellion, and Laura Schiavo, for example, consider the cultural meanings of perspective and representation in the eighteenth and nineteenth centuries by focusing on the emergence of particular visual media (zograscopes, the physiognotrace, and stereoscopes, respectively) and discussing how such media influenced notions of individual identity. Patricia Crain, Katherine Stubbs, and Diane Umble, by contrast, consider the cultural meanings of communication by focusing on the arrival and adaptation of particular networked media (optical telegraphs, electric telegraphs, and telephones, respectively) that helped shape notions of identity in relation to larger communities.

All of the authors engage new media as evolving, contingent, discursive frames, sites where the unspoken rules by which Westerners know and enjoy their world are fashioned. Such "rules" continually change, as new media become situated and as such adjustments inevitably redraw the boundaries of communities, including some individuals, and excluding others. Each new medium in effect helps to produce a distinct public. Erin Blake's work on zograscopes, for example, elaborates the idea that media assist in the construction of the modern, Western public sphere, with its corresponding liberal subject (today known as "the consumer"). Although she draws upon the work of Jürgen Habermas, Blake ignores the often-mentioned circulation of print media as the basis of the public sphere, instead looking to shared social practices to understand how space is visualized. Her public is literally a sphere; in her essay the bourgeois circles of eighteenth-century London pop into 3-D as they enter the rational and impersonal arena of public space via engravings glimpsed through new optical devices. This new medium, according to Blake, helps the public to map itself. Wendy Bellion's work on the physiognotrace depicts an American public that also maps itself, but this public is one more complicated by its own experiences of both graphic and political self-representation. By analyzing the American reception of this profile or silhouette-tracing device, Bellion introduces her readers to the cartography of the public sphere, showing the ways in which new media are adapted within the very discursive conditions, the very rules that they help to transform.

The rules for inclusion, for drawing the boundaries of a public sphere, are less concealed in Patricia Crain's essay, which examines how elaborate pedagogical systems designed to resemble new media interpellated and located their subjects, in this case by making them perform as optical telegraphs within larger, oppressive systems of cultural replication. Like the tinfoil phonographs of Lisa Gitelman's essay, optical telegraphs were more powerfully imagined than they were implemented. Very few were ever built or used, yet the idea of them circulated widely within the mentality, the public imagination, of their age. Joseph Lancaster's classroom telegraphs literally disciplined students, while even broader disciplinary measures may be read in their controlling institutional contexts, as well as glimpsed in the titles of early American newspapers like the *American Telegraph* [Conn.], *Hillsboro* [N.H.] *Telegraph*, and *Lincoln* [Me.] *Telegraph*. (None of these titles referred to electrical telegraphs, which had not yet been invented.) In Benedict Anderson's formulation, the circulation and ritualized consumption of newspapers like these assisted in the imagination of a national community. What their titles and Lancaster's system suggest, according to Crain, is that the imagination of *media* conditioned the imagination of communities. Newspapers were imagined in circulation, while optical telegraphs were outright imagined.

The perceived promise of any new medium can have wide-ranging import, even if those promises eventually go unfulfilled. To many observers, the tinfoil phonographs of 1878 promised a new, more modern and immediate type of text, as recordings might indelibly "capture" speech, without the intercession of literate humans wielding pencils and paper. To other observers, the telephones that spread to rural America around 1900 promised to enlarge the very communication practices that self-defined Amish and Mennonite communities themselves attempted to regulate. The wide popular reception of the first promise, Lisa Gitelman speculates, challenged and helped to transform vernacular experiences of writing and print, while raising questions about the instruments and the subjects of public memory. The Old Order perception of the second promise, Diane Umble shows, helped divide the aggregate Amish and Mennonite population, for this perception coincided with the ongoing regulation of intra- and inter-group communication and excommunication. Although so often the focus of great attention and optimism, new media are not, as these authors pointedly demonstrate, inherently benign; they "bite back."[10] They thrive amid unforeseen consequences, often despite the best, most vigorous intentions of their inventors, their promoters, their initial consumers, or of the customary arbiters of public intelligence.

Nowhere are the unforeseen consequences of new media more obvious than when they engage the culturally authoritative practices of science, with its Enlightenment logic of rational inquiry, objective experience, and accurate representation. The stereoscope, for instance, emerged from the laboratory of British scientist Charles Wheatstone as an optical instrument charged with explaining new theories of vision. To scientists, the stereoscope could be used objectively to demonstrate that vision is subjective, that the body can produce its own experiences of depth when presented with the right cues. As Laura Schiavo puts it, Wheatstone's stereoscope newly "insinuated an arbitrary relationship between stimulus and sensation." Yet within the context of commercially exploited and popularly apprehended photography, stereoscopes were ultimately recast as mimetic amusements that tendered to consumers an instructive and positivist model of how their eyes actually worked to see the world as it really is. Vernacular discourse, in other words, completely inverted the meaning of what the stereoscope "proved." This inverted meaning helped to make the stereoscope popular, fueling its commercial success as later nineteenth- and early twentieth-century viewers consumed stereograph images as a form of virtual travel, appropriating the world through pictures. At stake was far more than the prestige of Wheatstone or the anti-intellectualism of the marketplace. The rules by which the West knew the world had again come into play. The popularity of stereoscopes helped redraw the very category of the "real," the consensual practices of "accurate" representation.

Assumptions about what count as "rules," about what is "real" or "accurate" or "normal," are no less at issue when new media are less popular than stereoscopes were or less patently involved in describing normal human perception. Media emerge and exist in ways that both challenge and regulate notions of what it means to be human. Gregory Radick's essay provides a clear-cut case. An amateur ethologist using the new medium of recorded sound set out to learn the "language" of monkeys and stumbled into one of the hottest debates in Victorian evolutionary biology and linguistics: How is language uniquely human? In the course of his research, Richard Garner's recording phonograph became an instrument of knowledge deployed in various philosophic and scientific controversies—in the tension between amateur and professional science, for example, or in the dispute over whether abstraction or instinct founds thought and language, or in discussions about the fundamental differences between humans and animals. Garner worked on monkeys, but not without meddling with the category of the human in two ways. First, he raised anew the definition of "Man" as "the talking animal"; second, he wielded his phonograph as if it were a necessary—and better—third ear.

As Garner's third ear suggests—and as many authors have noted—new media can be viewed as an endeavor to improve on human capabilities. Like a telescope added to the eye of an astronomer or a microscope added to the eye of a biologist, media can extend the body and its senses. Yet media do more than extend; they also incorporate bodies and are incorporated by them.[11] Media are designed to fit the human, the way telephone handsets or headsets literally fit from ear to mouth, but also the way telephone circuits, satellites, and antennas fit among their potential consumers, as integral parts of communication/information networks that literally shape what communication entails for individuals in the modern age. And if media fit humans, humans adjust themselves in various ways to fit media, knowingly and not. Hands physically adjust themselves to different keyboards, different keypads, and different pointing devices, while users subtly adjust their sense of who they are. Some of these complexities may be glimpsed in Katherine Stubbs's essay, which reads the history of electrical telegraphy in the United States against and within the fiction that appeared in telegraph trade journals. Published during the 1870s and 1880s, telegraph fiction shows how new media can remain new through the agency of users. Amid ongoing conflicts between labor and capital arising in part from the feminization of the workforce, telegrapher-authors both used and represented the telegraph as a means to explore identity in its relation to the body. In remaking themselves, by negotiating gender-at-a-distance-and-by-telegraph, for instance, telegraphers kept the character of their medium unsettled. In other words, the "newness" of new media is more than diachronic, more than just a chunk of history, a passing phase; it is relative to the "oldness" of old media in a number of different ways.

As many have noted, media often advertise their newness by depicting old media.[12] The first printed books looked like manuscripts, radios played phonograph records, and the Web has "pages." Ellen Gruber Garvey and Paul Young each explore less familiar instances in which the new represents the old in order to understand more fully the purchase that "newness" has on the process of representation. As Garvey's account of scrapbooks explains, scrapbook-makers took old media—literally the old books and periodicals they had lying around—and made them into new media in the form of scrapbooks. "Newness" in this case resonated as much with personal and domestic experiences as it did with public and collective apprehensions of novelty, posterity, or periodicity. Scrapbook-makers tampered with the meanings of the scraps they collected by collecting them, a practice Garvey refers to as "gleaning" and connects to the composition and use of the Web today. Young, on the other hand, presents a "telegraphic history of early American cinema," reading filmic representations of telegraphs as only the most obvious link between these two media, which seem, in retrospect, so different. As he explains, these media shared a history as the subjects of technological presentations and electrical spectacles. From the start, both became instruments of news reportage, one in the transmission of stories on the wire (that is, by telegraph wire services like the Associated Press) and the other in the projection of stories onto the screen in "actualities" and proto-newsreels. "Newness" in this case resonated with emergent conventions for representing narrative time, with experiences of currency (of news as either new or old), and with new technology—all experiences that transform our sense of time and space.

We hope these essays will help to broaden the inquiry of media studies by calling attention to the ways media are experienced and studied as the subjects of history. No ten essays can do more than open the question, but opening the question is crucial, we think, particularly as today's new media are peddled and saluted as the ultimate, the end of media history. "Newness" deserves a closer look. To that end, we include a brief section of documents for discussion. These documents are not illustrations of our text as much as they are artifacts that themselves point toward the rich and diverse record available to media historians. We hope that they will suggest specific historical and cultural meanings for media and promote a broader discussion of media history. Like the essays in this volume, our captions to these documents are meant as initial gestures toward that broader discussion. We include them to remind readers that the history of media is an ongoing, highly self-reflexive conversation about what we mean and—literally—*how* we mean it.

Notes

1. Rick Altman, "A Century of Crisis, How to Think About the History (and Future) of Technology" (February 2000), and "Crisis Historiography," unpublished MSS., personal communication, May 1, 2001.

2. Carolyn Marvin, *When Old Technologies Were New: Thinking about Electric Communication in the Late Nineteenth Century* (New York: Oxford University Press, 1988).

3. Deadmedia.org. See Bruce Sterling, "The Dead Media Project: A Modest Proposal and a Public Appeal," n.d., <http://www.deadmedia.org/modest-proposal.html>, June 2001.

4. "Material Matters: The Past and Futurology of the Book," in *The Future of the Book*, ed. Geoffrey Nunberg (Berkeley: University of California Press, 1996), 65.

5. Timothy Lenoir, "Inscription Practices and Materialities of Communication," in *Inscribing Science: Scientific Texts and the Materiality of Communication*, ed. Timothy Lenoir, 1–19 (Stanford: Stanford University Press, 1998), 7–8.

6. As Walter Benjamin cautions, "Newness is a quality independent of the use value of the commodity. It is the origin of the illusory appearance that belongs inalienably to images produced by the collective unconscious. It is the quintessence of that false consciousness whose indefatigable

agent is fashion. This semblance of the new is reflected, like one mirror in another, in the semblance of the ever recurrent. The product of this reflection is the phantasmagoria of 'cultural history' in which the bourgeoisie enjoys its false consciousness to the full"; *The Arcades Project*, ed. Howard Eiland and Kevin McLaughlin, trans., Rolf Tiedemann (Cambridge: Harvard University Press, 1999), 11.

7. James W. Carey, *Communication as Culture: Essays on Media and Society* (Boston: Unwin Hyman, 1988).

8. Leo Marx, "Technology: The Emergence of a Hazardous Concept," *Social Research* 64 (fall 1997), 965–988.

9. Naomi Mezey, "Law as Culture," *Yale Journal of Law & the Humanities* 13, no. 1 (2001): 35. See also Raymond Williams, *Keywords: A Vocabulary of Culture and Society* (New York: Oxford University Press, 1985).

10. "Bite back" is from Edward Tenner's title, *Why Things Bite Back: Technology and the Revenge of Unintended Consequences* (New York: Alfred A. Knopf, 1996).

11. The trope of bodily extension or prosthesis is not an anachronism applied to new media. As James Lastra shows, it is one of two tropes that have played a normalizing role in the emergence of modern media (the other is that of inscription); *Sound Technology and the American Cinema: Perception, Representation, Modernity* (New York: Columbia University Press, 2000); see "Introduction" and chapter 1. See also N. Katherine Hayles on "incorporating practices and embodied knowledge," 199–207 in *How We Became Posthuman: Virtual Bodies in Cybernetics, Literature, and Informatics* (Chicago: University of Chicago, 1999).

12. The remediation of one medium by another newer medium has recently been explored by Jay David Bolter and Richard Grusin in *Remediation: Understanding New Media* (Cambridge: MIT Press, 1999). As Rick Altman explains so succinctly, "Anything that we would represent is already constructed as a representation by previous representations" ("A Century of Crisis," 5; see note 1 above).

Documents

JANUARY, Second Winter Month.

Days.	☉	♄	♃	♂	♀	☿
1	♑ 11 42	♏ 26	♈ 2	♑ 16	♒ 28	♐ 21
7	17 49	26	3	21	♓ 5	24
13	22 41	26	3	26	11	♑ ½
19	29 48	26	4	♒ 1 ½	17	7
25	♒ 5 54	26	5	5	22	15

Of the SUN.

MR. WHISTON fays the Sun is 763,000 miles in diameter, and is 230,000 times bigger than the Earth, and is eighty-one millions of ftatute miles diftance from the Earth, each of which miles is 5280 Englifh feet.

Of the EARTH.

THE Earth, according to Mr. Whifton, is 7970 miles in diameter ; which will make nigh 24000 miles in circumference. It revolves about its Axis in 23 hours 56 minutes : It moves in the fpace of one hour 56,000 miles, and is 365 days 6 hours and 9 minutes revolving about the Sun.

Of the Divifibility of MATTER·

THE ingenious Mr. Lewenhoek fays, that in the Milt of one Cod-Fifh there are more little animals, than there are inhabitants on the face of the earth. Several thoufands of them can ftand upon a needle's point. And Dr. Keil has fhewn, that the fmalleft vifible grain of fand would contain more of the Globules of that fluid (which ferves thefe animals for blood) than ten thoufand two hundred and fifty-fix of the higheft mountains in the world would contain grains of fand. *A damd lye*

Of the Fixed STARS.

DR. HOOK and Mr. Flamfteed fay, that the diftance of the fixed Stars from the Sun is fo great, that a bullet fhot out of a mufket would not reach them in 5000 years. *Another*

Document A. Almanac Page with Reader's Annotations (1774) This page from Gleason's *Massachusetts Calendar* contains a reader's annotations, "A damd lye [sic]," and "Another," beside accounts of Anthony van Leeuwenhoek's microscopy and John Flamsteed's astronomy. The reader's skepticism calls attention to the complicated economies of belief that engaged eighteenth-century readers. While almanacs like Gleason's possessed defining predictive functions, they also appealed to readers' sense of irony and doubt. What the reader doubted in this case is science, though, not fanciful weather forecasts or astrology. Nature is itself newly and doubly mediated, first by the century's fledgling optics—microscopes and telescopes—and second by the printed page. Courtesy, American Antiquarian Society.

Document B. Image Sellers from Published Street Cries (1810–17)   Street cries were a popular print genre at the end of the seventeenth century, adapted for children at the beginning of the nineteenth century. They offer an interesting way to think about the character of literacy and the functions of media. As instructional texts, the street cries promoted literacy, paradoxically, by romanticizing the oral, purporting to represent the appearance and repeated cries of itinerant peddlers in the cities of Western Europe and America. They gesture toward communities defined by earshot and integrated by the circulation of produce, goods, and services in the hands of outsiders, while they domesticate the peddler's voice by offering it for reiteration in the child's pronunciation of the alphabet. In these examples images circulate as commodities in the hands—and from the lips—of an Italian, while woodcut images (of images) circulated in the hands of publishers, parents, and children who wanted to connect the name and the image of the letter "I" with its phonetic identity. Courtesy, American Antiquarian Society.

Document B *continued*

Document B *continued*

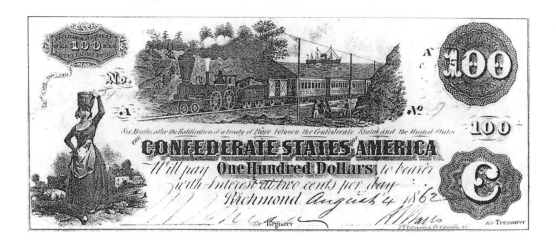

Document C. Banknotes (1852, 1862)   Before a uniform federal currency was established in the United States during the Civil War, more than a thousand individually state-chartered banks issued their own paper money. Banknotes were an intensely complicated form of media. They circulated locally, regionally, and nationally, as they fluctuated in worth according to perceived and publicly reported redemption values. A wide variety of different engravings on notes helped to make them look valuable, or look real, though the duplication and reuse of images was common for legitimate as well as for forged notes. In these examples, the circulated notes with their recirculated images doubly represent the circulation of capital: the bills were supposedly backed by gold, even as they are fronted by the same tiny locomotive (with the same tiny telegraph line beside it) and the same tiny steamship, allegories of the dominant and intensively capitalized transportation networks of the day. Smithsonian Institution, National Numismatic Collection.

COMPARATIVE TIME-TABLE.

Showing the Time at the Principal Cities of the United States, compared with Noon at Washington, D. C.

There is no "Standard Railroad Time" in the United States or Canada but each railroad company adopts independently the time of its own locality, or of that place at which its principal office is situated. The inconvenience of such a system, if system it can be called, must be apparent to all, but is most annoying to persons strangers to the fact. From this cause many miscalculations and misconnections have arisen, which not unfrequently have been of serious consequence to individuals, and have, as a matter of course, brought into disrepute all Railroad Guides, which of necessity give the local times. In order to relieve, in some degree, this anomaly in American railroading, we present the following table of local time, compared with that of Washington, D. C. :

NOON AT WASHINGTON.	NOON AT WASHINGTON.	NOON AT WASHINGTON.
Albany, N. Y............12 14 P.M.	Jackson, Miss............11 08 A.M.	Petersburg, Va..........11 50 A.M.
Augusta, Ga.............11 41 A.M.	Jefferson, Mo...........11 00 "	Philadelphia, Pa........12 08 P.M
Augusta, Me.............12 31 P.M.	Kingston, Can...........12 02 P.M.	Pittsburg, Pa...........11 48 A.M.
Baltimore, Md...........12 02 "	Knoxville, Tenn.........11 33 A.M.	Plattsburg, N. Y........12 15 P.M.
Beaufort, S. C. ....11 47 A.M.	Lancaster, Pa...........12 03 P.M.	Portland, Me............12 23 "
Boston, Mass...........12 24 P.M.	Lexington, Ky...........11 31 A.M.	Portsmouth, N.H.........12 25 "
Bridgport, Ct...........12 16 "	Little Rock, Ark........11 00 "	Providence, R. I........12 23 "
Buffalo, N. Y...........11 53 A.M.	Louisville, Ky..........11 26 "	Quebec, Can.............12 23 "
Burlington, N. J........12 09 P.M.	Lowell, Mass,...........12 23 P.M.	Racine, Wis.............11 18 A.M.
Burlington, Vt..........12 16 "	Lynchburg, Va...........11 51 A.M.	Raleigh, N. C...........11 53 "
Canandaigua............11 59 A.M.	Middletown, Ct..........12 18 P.M.	Richmond, Va............11 58 "
Charleston, S. C........11 49 "	Milledgeville, Ga.......11 35 A.M.	Rochester, N. Y.........11 57 "
Chicago, Ill............11 18 "	Milwaukee, Wis..........11 17 "	St. Louis, Mo...........11 07 "
Cincinnati, O...........11 31 "	Mobile. Ala.............11 16 "	St. Paul, Min...........10 56 "
Columbia, S. C..........11 44 "	Montpelier, Vt..........12 18 P.M.	Sacramento, Cal.........9 02 "
Columbus, O.............11 36 "	Montreal, Can...........12 14 "	Salem, Mass.............12 26 P.M.
Concord, N. H...........12 23 P.M.	Nashville, Tenn.........11 21 A.M.	Savannah, Ga............11 44 A.M.
Dayton, O...............11 32 A.M.	Natchez, Miss...........11 03 "	Springfield, Mass.......12 18 P.M.
Detroit, Mich...........11 36 "	Newark, N. J............12 11 P.M.	Tallahassee, Fla........11 30 A.M.
Dover, Del..............12 06 P.M.	New Bedford,............12 25 "	Toronto, Can............11 51 "
Dover, N. H.............12 37 "	Newburg, N. Y...........12 12 "	Trenton, N. J...........12 10 P.M.
Eastport, Me............12 41 "	Newcastle, Del..........12 06 "	Troy, N. Y..............12 14 "
Frankfort, Ky...........11 30 A.M.	New Haven, Ct...........12 17 "	Tuscaloosa, Ala.........11 18 A.M.
Fredericksburg..........11 58 "	New Orleans, La.........11 08 A.M.	Utica, N. Y.............12 08 P.M.
Galveston, Texas........10 49 "	Newport, R. I...........12 23 P.M.	Vincennes, Ind..........11 19 A.M.
Halifax, N. S...........12 54 P.M.	New York, N.Y...........12 12 "	Wheeling, Va............11 45 "
Harrisburg, Pa..........12 01 "	Norfolk, Va.............12 03 "	Wilmington, Del.........12 06 P.M.
Hartford, Ct............12 18 "	Northampton,Ms..........12 18 "	Wilmington, N. C........11 53 "
Huntsville, Ala.........11 21 A.M.	Norwich, Ct.............12 20 "	Worcester, Mass.........12 21 P.M.
Indianapolis, Ind.......11 26 "	Pensacola, Fla..........11 20 A.M.	York, Pa................12 02 "

By an easy calculation, the difference in time between the several places abovenamed may be ascertained. Thus, for instance, the difference in time between New York and Cincinnati may be ascertained by simple comparison, that of the first having the Washington noon at 12 12 p. m., and of the latter at 11 31 a. m ; and hence the difference is 41 minutes, or, in other words, the noon at New York will be 11 19 a. m. at Cincinnati, and the noon at Cincinnati, will be 12 41 at New York.

Document D. Comparative Timetable (1868) Originally printed in the *Travelers' Official Railway Guide of the United States* for 1868, this timetable raises the question of standards pertinent to all new media systems and debunks the widely held notion that telegraphy caused the standardization of "railroad time" and eventually the creation of global time zones. The schedule presents a snapshot of a forgotten world, one with telegraphs but not telephones, in which movement across extended geographical space was an experience shaped by regionally dominant service providers (that is, the railroads) and by localized sources of orientation. From such documents standardization emerges as a process, not an effect, one dependent as much upon changing corporate structures and human experiences as upon any single technological device. *Official Railway Guide*, Commonwealth Business Media, Inc.

BURPEE'S SEED WAREHOUSE,

located at Nos. 475 and 477 North Fifth Street, extending through to Nos. 476 and 478 York Avenue, is the largest in the city of Philadelphia; and, for the conducting of an extensive mail seed business, is pronounced, by all who have examined the various departments, to be the most admirably arranged in this country. It occupies four building lots; is five stories high, besides basement, and has a wide frontage on both Fifth Street and York Avenue—both broad streets—thus giving ample light and ventilation. Customers visiting Philadelphia are cordially invited to call, and can reach our warehouse conveniently from any station; the Fifth Street cars pass every minute or two. That our new customers at a distance may appreciate our facilities for the *prompt* and *careful execution* of orders, we give illustrations of some of the departments, which we will briefly describe.

THE MAIL ORDER DEPARTMENT, shown above, occupies the entire second floor, and is devoted entirely to the filling of orders for seeds, to be sent, postpaid, by mail, and to the Flower Seed Department. Here are filled several thousand orders a day in winter and spring. The seeds are already done up, *by the million of packages*, in packets, ounces, quarter pounds, pounds, pints and quarts, and can readily and quickly be selected on order, from the thousands of compartments, drawers and bins, all distinctly labeled and numbered. After the orders are filled, they are carefully examined by expert employés, whose entire time is devoted to this work, to see that no mistake has occurred; they are then passed to men standing at large tables, to be wrapped and tied, when they are taken, by the elevator, to the stamping room, on the first floor, and thence carted, in sacks, to the post office.

OPENING THE MAIL. The mail office is the scene of quiet but busy work, the entire time of several persons being required to open and assort the mail; by continued practice they have become very expert, as may be judged from the fact that, in the busy season, *our mail numbers three to five thousand letters a day.*

THE BUSINESS OFFICE, of which the illustration only gives a partial view, is located on the first floor, entrance from Fifth Street. Here the books are kept, and the orders received are arranged on reversible files, with the address tag or label attached to each order, and are then recorded in order books, of which there is a separate book for every State in the Union. Here, also, most of the correspondence is conducted. All the office work is completed on the orders the day they are received; they are then ready to be filled the day following, and the manager of each department thus knows early in the morning exactly what amount of work he has to map out for the employés.

THE FLOWER SEED DEPARTMENT, under separate management, with experienced assistants. Here all orders for Flower Seeds and Bulbs are filled. The manager having had twenty years' experience in this department of the seed business, accuracy and care are assured. ☞ *See pages 84 to 95 for Novelties, and pages 96 to 115 for* General List of Flower Seeds.

Document E. Seed Catalog (1888) Though they are frequently overlooked in favor of mail order houses like Montgomery Ward and Sears, Roebuck, seed companies early adopted distance-selling, and they offer an interesting way to think about the history of information systems. As these pages of *Burpee's Farm Annual* reveal, seed sellers tried to integrate the circulation of mail, of revenue, and of genetic information into a single, most

EXPRESS AND FREIGHT ORDER DEPARTMENT.

THE EXPRESS AND FREIGHT ORDER DEPARTMENT, shown above, is located on the third floor, and is devoted exclusively to the packing of seeds that go by freight or express. Here is kept in large drawers, bins, barrels and in sacks, a complete duplicate stock of all the garden and farm seeds catalogued by us—all flower seeds being obtained, on order, from the floor below. Only men are employed in this department, and by years of practice they have become very efficient in packing boxes, barrels and sacks of seed neatly and securely. As soon as filled, the orders, each accompanied by a memorandum sheet, specifying number and description of packages, route, etc., and signed by the "filler," are sent by the elevator to the ground floor, where the shipping clerk makes out the bills of lading, and sees that the wagons are properly loaded for the different freight stations. Each customer is then advised, by mail, of the shipment. The elevator shaft is located about twenty-five feet back from the northern door on the York Avenue side; thus a wide, unobstructed area, extending to the wide doorway south, allows ample room for the sorting of the sacks, boxes and barrels intended for the different railroads.

THE SHIPPING OFFICE is located south of the southern door on York Avenue. In this office the shipping clerk and his assistant, besides making out bills of lading and records as to shipments, also notify the consignees of the shipment of each order.

A STORE ROOM, where is kept a portion of our stock of Peas, Beans, Sweet Corn and Small Garden Seeds, in bags, the bulk of the seed gram being stored on other floors and in another warehouse.

SEED POTATO STORES. Our cellar is specially adapted for the safe storage of a large stock of Seed Potatoes, which we receive each year from our growers in the North, in the fall, and keep for the use of our customers in the spring.

AMONG OTHER DEPARTMENTS which are not illustrated, we might mention The Private Office, which is placed in communication with different parts of the building by means of electric bells and speaking tubes, The Stamping Room, The Stock Seed Room, The Seed Cleaning Room, and The Printing Office.

It is our rule to fill all orders the day after they are received. This we are enabled to do by our excellent facilities and large force of skilled employés, many of them practical seedsmen of long experience. Our *promptness* in filling orders, together with the quality of our seeds, has gained us such a host of friends that we have the largest mail trade in Philadelphia, and the largest mail seed business of the kind in the United States.

WE GUARANTEE that all seeds and other goods ordered of us shall reach the purchaser safely and in good condition. We are also responsible for the safe receipt of all money sent to us, if remittances be made by P. O. order, registered letter, express order, or bank draft. Thus, even customers in the most distant States and Territories can obtain their supplies as safely and have their orders as carefully executed as if they had called in person at our warehouse.

W. ATLEE BURPEE & CO.

SHIPPING OFFICE

A STORE ROOM

SEED POTATO STORES

efficient system. The Burpee company wanted no random broadcast. (All of the original uses of the term "broadcasting" dealt with spreading seeds.) Seeds needed to be collected and saved as well as sold and mailed. Although their purpose was utilitarian, seed companies sold many ornamental varieties and "novelties" as well. Smithsonian Institution, Horticulture Branch Library.

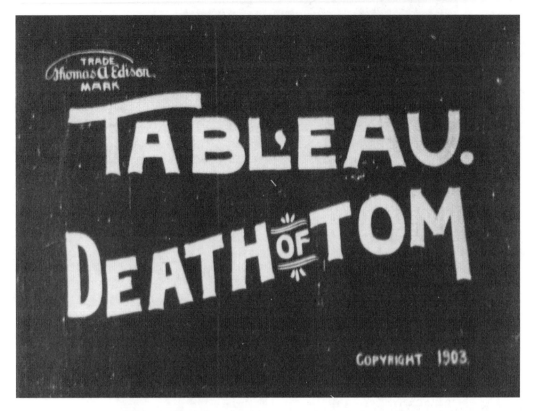

Document F. Film Intertitle (1903)   *Uncle Tom's Cabin* was the first motion picture to use intertitles. This title announces the final scene of the film, and it suggests that the new medium of moving pictures was curiously entangled with theatrical conventions (tableaux are static, not moving) while at the same time engaged with innovations in narrative strategies, intertitles among them. Catalog copy assured exhibitors that "every scene [has been] posed in accordance with the famous author's version," though the copyright announced in this intertitle belonged to the Edison Manufacturing Company, not the novelist Harriet Beecher Stowe. The "author's version" was common property by 1903, and film audiences would have known the story and its characters even without the intertitles. Library of Congress.

Document G. Newspaper Comic Strip (1907) Comic strips were new at the end of the nineteenth and beginning of the twentieth centuries, when big-city newspapers began to capitalize on artists like R. F. Outcault and Winsor McCay in order to attract readers. This McCay example from April 7, 1907, helps illustrate the raucous intertextual quality that in part defined the form. The strips made sense among and against other media of the day (there had already been a movie version of *The Rarebit Fiend* in 1903), including the columns of newsprint amid which they appeared. Unlike the news, these strips were panels rather than columns, more visual and less literate, and they were motivated by a distinct narrative logic and constructed a different version of the day's events. This was particularly true of McCay's continuing dream-fantasy strips, *Little Nemo in Slumberland* and *Dream of the Rarebit Fiend*. Here the self-referring subject of McCay's strip recognizes a breakdown in communication as inkblots corrupt an ideal, less mediated, congress between the artist's mind and the subject's body, or between the day's fashions and their representation in the newspaper. Permission of R. Winsor Moniz.

Document H. Backgrounds and Headgrounds for Photographers (1908) This selection from the Sears, Roebuck catalog addresses the history and social functions of photography while raising a number of issues about how images were produced and consumed across media. The sale of painted backgrounds for portrait photographs recalls the meaning of photography before Eastman Kodak created a market for amateur, everyday picture taking, starting in 1888. This catalog copy suggests the wide circulation of different and uniquely personal portraits, all with identical, impersonal backgrounds. Documents like it can contribute precision to discussions of "the age of mechanical reproduction," at least by undermining assumptions that the distinction between painting and photography is always one between uniqueness and mass production, or that painting and photography always form antithetical sources of meaning.

1 Zograscopes, Virtual Reality, and the Mapping of Polite Society in Eighteenth-Century England

Erin C. Blake

Although primarily associated today with B movies and IMAX headgear, 3-D viewing technology has a very long history. As early as 1677, German writer Johann Christoph Kohlhans described how equipping a common camera obscura with a convex lens gave a pictured object the appearance of being "before one's naked eye in width, breadth, familiarity and distance." By 1692, William Molyneux of Dublin could write as if this three-dimensional effect were already well known. "Pieces of Perspective," he said of architectural views, "appear very Natural and strong through Convex Glasses duly apply'd."[1] Yet despite the vivid description by these students of optics, contemporary instrument sellers did not promote the use of convex lenses for picture-viewing, and print sellers did nothing to advertise perspective views as particularly suited for use with lenses. The 3-D phenomenon was known across Europe, but it was neither popularly disseminated nor commercially exploited.

All of this changed in the 1740s when convex lenses and perspective prints became packaged together as consumer goods in England. The first 3-D gadget craze developed around a manufactured lenticular device commonly referred to today as the zograscope. Between the mid-1740s and the mid-1750s, zograscopes and zograscope prints appeared regularly in English magazine copy and newspaper advertisements, as did hundreds of different engraved, hand-colored images designed for use with the device.[2] Curiously, almost every one of the known engravings from that period has the same subject. Zograscope prints depict the manmade environment, particularly urban topography.

Why was this early incarnation of three-dimensional viewing so relentlessly invested in cosmopolitan views—in cities, buildings, and gardens—rather than in natural scenery, or the wide range of subjects tapped by painting and other graphic genres? What made the relationship between the public subject matter and the domestic practice of zograscopic viewing so attractive? The answer lies with the social function of the device: The zograscope enabled its users to think of themselves as individuals participating together in the larger sphere of polite society. To the newspapers, correspondence, novels, and conversation commonly held up as instruments of polite society, the zograscope offers a crucial nonverbal complement. By creating a three-dimensional space for viewers that was both domestic and public, zograscopes allowed users intricately to develop a new relationship between privacy and the public sphere. Zograscopes provided a model for seeing public space as generic, neutral "polite" space.

By the 1760s, although they were still being sold in large numbers in England, zograscope views fell out of the spotlight. They had gone from being a topic of discussion in the previous decade to being an ordinary amusement not worthy of polite society's comment. It was around that time that the production and consumption of zograscope images seem to have developed in France, Germany, and Italy, but never with the same brief intensity and limited subject matter as they had first had in England.[3]

The Viewing Device

The essential component of a zograscope is a convex lens at least three inches in diameter, with a focal distance of about an arm's length. The lens refracts the light rays coming from each point on the print being viewed so that rather than seeming to originate on the surface of the paper, they arrive at the eyes almost parallel to each other, as if originating from a much greater distance. Spherical aberration around the edges of the lens and juxtaposed patches of bright color on the print itself further disrupt normal depth cues. These disruptions combine to create the perception of three-dimensional picture space known as *stereopsis*.[4] The single view stretches into depth, with the individual figures, carriages, and trees in the image still flat, but apparently distanced from each other.

In the eighteenth century, the lens of the zograscope was usually suspended vertically in front of a mirror that was inclined at a 45-degree angle, and the whole assembly either mounted on a stand or inside a box that could be placed upon a parlor table (figure 1.1). Instead of looking through the lens directly at a print, the viewer looked at the reflection (in the flat mirror) of the refraction (through the curved lens). The fact that the scene and caption would appear in reverse with this arrangement appears not to have bothered viewers in England, at least not enough for deliberately reversed prints (with reversed captions) to be produced in great numbers, as they were somewhat later in France, Germany, Italy, and elsewhere. Once placed before the zograscope, the earliest three-dimensional images left their captions behind.

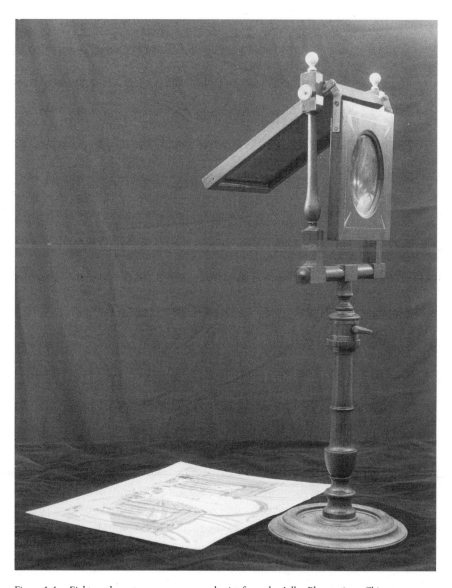

Figure 1.1   Eighteenth-century zograscope and print from the Adler Planetarium, Chicago.

Virtual Reality and "Polite Society"

Above all, eighteenth-century sources emphasized the lifelikeness of zograscope views. William Emerson praised the zograscope for its ability to put viewers inside a summer scene, even on a winter's day. Henrietta Knight, Lady Luxborough, felt the same way, and she wrote in 1748 that she preferred experiencing London with her zograscope, because it made the place come to life without her having actually to visit a city of which she was not very fond. Both Knight and Emerson, then, were describing the visual aspects of what is today called "virtual reality," a space that is enacted technologically and perceived by viewers but that exists nowhere on its own.[5] The term *virtual* describes something intangible but real in its effect, as in "virtual image" in optics, "virtual memory" in computing, and in keeping with Wolfgang Iser's use of the term, the "virtual" literary work produced when reader and text converge. The zograscope view does not exist on a piece of paper or on the lens, but is a simulacrum that only comes into being in the viewer's visual cortex.

Like the virtual image produced by the nineteenth-century stereoscope, the zograscope view does not have an original elsewhere; it is created then and there, and exists only in the moment that viewer and viewing device come together. This places the zograscope ahead of its time according to Jonathan Crary's chronology in *Techniques of the Observer,* where a purely mechanical "camera obscura model" of vision characterizes the eighteenth century.[6] Not until early nineteenth-century discoveries about persistence of vision and other physiological elements of optics, he writes, was it possible for viewers to consider a virtual image as a form of reality rather than as an irrelevant deception. For Crary, the stereoscope exemplifies this new conception of vision. Should his model be pushed back in time so that the optically similar zograscope can play the hero instead? Tempting as it is, the answer is no. The zograscope of 1750 did not provide a new dominant model of seeing; it provided a different way for a limited group of people, the men and women who considered themselves members of polite society, to visualize public space.

The imagined (but not imaginary) community of polite society knew no physical boundary. By self-definition it was not confined to particular neighborhoods, cities, or countries, but rather existed within a conceptual boundary as the group of relatively well off, educated, decorous people of good character. Much has been written in recent years about the emergence of this category, its independence from monarchical or ecclesiastical structures, and the difficulty of assigning it a precise boundary and name.[7] What I call polite society here would be termed the "public sphere constituted by private people" in Jürgen Habermas's canonical text.[8] I prefer the word "polite" because it was used by contemporaries to set themselves apart, but unlike "the Republic of Letters" (also used at the time) it does not imply a verbal medium for doing so.[9] Newspapers, correspondence, conversation, art criticism, and novels are often cited as means through which people came to think of themselves as individuals participating together in polite society. The zograscope provided a uniquely visual guide for the same subjectivity.

Zograscope views provided a spatial understanding in which space can be theorized as a neutral thing, apart from discourses of authenticity or narrative. By creating a virtual space that was domestic and public at the same time, zograscope views allowed observers to imagine a new relationship with the nondomestic world around them. Bodily experience grounded in the fullness of sights, sounds, smells, and jostling in the streets was too earthy and all-inclusive for a self-conception as polite society. Instead, zograscope views told a story of public space as available, accessible, dynamic, and vibrant, but controllable, clean, and polite. In this story, streets, squares, parks, and church interiors were spaces for unhindered physical movement and expansive vision, not static, deep, particular experience. It was a new modern geography,[10] made visible by the zograscope in its presentation of generalized open space as something three-dimensionally real.

A New Kind of Print

Advertisements reveal that before the late 1740s, publishers conceived of prints in three ways. At one end of the spectrum, they had value as collectible items for connoisseurs, where they would be stored in portfolios to be admired and discussed for their subject matter and artistic merits. In a more utilitarian way, prints could serve as wall decoration, either framed behind glass, or pasted on walls with decorative borders as a kind of wallpaper. Lastly, some prints were little more than amusing throwaways: souvenirs of a public event, comic scenes, miscellaneous designs for children, and so on, which peddlers might buy in bulk from a map- and printseller to sell again at fairs or in the streets. Zograscope views appeared on the scene as a fourth type of print, one that was not meant for connoisseurs' portfolios, nor to be hung on a wall, nor for children to play with. Zograscope prints, instead, were spectacular or "curious" scenes for adults. Not fine enough to collect as art, but not cheap enough to be children's playthings, they still do not fall into the remaining category of wall decoration. A Mr. Mear made a similar point when he divided his customers into three groups by "Taste," those who "want Prints for Furniture," those who want prints "to colour for the Diagonal and Concave Glasses," and those "who are forming a Collection or Cabinet of Prints."[11]

The additional explanation zograscope prints received in advertisements attests to the newness of the device. Collectors did not need to be told how to store or display their fine-art prints, or what the experience of looking at them was like. The aesthetic value of fine-art prints for collectors was already understood. The zograscope experience, on the contrary, was unknown except perhaps to people interested in natural philosophy who had read optics treatises or attended public lectures. Hence, map- and printsellers included tempting tidbits in their advertisements, which surely affected observers' expectations. "These Prospects viewed in a Concave Glass have a surprising Appearance," wrote S. Austen and W. H. Toms, with three little stars setting off the phrase. Thomas Bakewell said his eight prints, when seen with a zograscope, "have a greater Effect than any Thing every [sic] yet done."[12]

The earliest mention of prints for use with a zograscope appeared in the *St. James's Evening Post* for 2 to 4 April 1747.[13] References to zograscope views increased rapidly after that. The number of newspaper advertisements peaked in 1749, the same year the *Gentleman's Magazine* described zograscope viewing. Then in 1750 and 1751, the *Gentleman's Magazine* explicitly included four zograscope views in each annual volume, saying in the list of plates that "Those marked with a Star, if coloured, will make very good Objects for the Optic Machine, or the Concave Mirror." By 1753, a stand-alone catalog of zograscope views had appeared, naming over two hundred prints in Robert Sayer's stock.[14]

Publisher John Bowles saw fit to invest in the new format and point of view for at least sixteen pictures of ostensibly the same subjects around 1750, even though his earlier plates were still perfectly usable, as demonstrated when twenty-two plates from his 1731 catalog appeared in the 1754–1755 edition of *Stow's Survey of London*.[15] A plate like the 1731 catalog entry *The Monument on Fishstreet-Hill, erected in commemoration of the Fire of London, 1666* (engraved by Sutton Nicholls), which is in portrait orientation, with banners in the sky, and the edifice itself blocking the recession into depth, could not be sold to consumers hungry for zograscope views (figure 1.2). John Bowles's new version of the plate, called *The Monument in memory of the conflagration 1666* in his 1753 catalog, is markedly different (figure 1.3). Oriented in landscape format, it has the standard 11 × 17-inch dimensions of a zograscope view, and the title now appears in the lower margin, where it will not interfere with the three-dimensional illusion. Most important, the visual emphasis has shifted away from the monument itself in order to draw attention to the swath cut into the picture space by the wide street, with its orderly parade of unremarkable traffic. A similar opening up of space can be seen in the other new Bowles plates.

Figure 1.2   Sutton Nicholls, *The Monument*, first published 1731, this impression 1754. Photo courtesy of The Newberry Library.

Robert Sayer, Mrs. Beckett, Mr. Watkins, and W. Harbet took care to reassure potential customers that their views, which could be used as wall decoration, were also "delightful for the concave and Diagonal Mirrours, and are allow'd by Connoissures to excel for either Purpose."[16] Clearly, true connoisseurs would neither need nor want a newspaper advertisement to instruct them in their purchases. Instead, the wording directs itself to people who aspired to a more cultured existence, and were willing to be convinced that a "Diagonal Mirrour," or zograscope, was part of the kit.

The Subject Matter

Although conversation pieces, idyllic rural scenes, portraits, and histories were all well represented in the visual arts of the period, they did not appear in commercial zograscope

Figure 1.3   Thomas Bowles, *The Monument of London in remembrance of the dreadfull Fire in 1666. Its Height is 202 feet.* | *Collone de Londres elevé pour une Perpetuelle Resouvenance de Lincendie Generalle de cette Ville en 1666. En Haut 202 Pieds,* circa 1750. Victoria and Albert Museum.

prints.[17] There is no technological reason this should be the case, since even a portrait would appear in 3-D with the device, as Benjamin Martin made clear when describing the effect a few years before the zograscope craze got underway. "The Cheeks are protuberant, the Nose projects, the Eyebrows overhang . . . and the Lips are about to speak," he wrote in 1740.[18] Nevertheless, only views were printed for the zograscope in England in the 1750s, and almost all of these views depicted places in and around the great cities of Europe.

Approximately half of the zograscope prints listed in John Bowles and Son's 1753 catalog showed scenes that are clearly urban, such as streets and rivers busy with traffic, groups of buildings, and cathedral interiors. The other half mostly showed landscaped gardens with geometrically laid-out walks populated by small well-dressed figures. They emphatically did not show sentimental rural scenes, bucolic peasants, or other "natural" types we might think of as the antithesis of urban. The so-called "parks" in this nonurban category were well-tended enclaves within London, like St. James's Park and Green Park, not vast hunting parks available only to aristocrats.

All two hundred and eighty-five zograscope prints advertised in London newspapers between 1747 and 1753 (after which time the number of new views fell dramatically) showed scenes in Western Europe.[19] British scenes dominated, with more views of London and environs than of the rest of Britain put together. Views of Italy came next, primarily Venice and Rome. The twenty-three plates in the 1750 volume of the *Gentleman's Magazine* show that the scarcity of other subjects in zograscope views was not because other subjects were beyond the realm of possibility for engravings at the time. The four plates in this volume, which "make very good Objects for the Optic Machine, or the Concave Mirrour," were typical of zograscope prints in that they showed modern places in Western Europe: two church interiors in London, a view of Fontainbleau (figure 1.4), and a view in Florence. The nonzograscope plates in the volume, however, showed a range of exoticism, erudition, and drama: African princes, antiquities, a snarling tiger, and an eruption of Mount Vesuvius engulfing whole towns in fire.

The generic sameness of the public, polite subject matter in zograscope views necessarily excluded specific images of Continental high culture, the experience of which was one of the ostensible points of a Grand Tour. The "Grand Duke's Gallery" in Florence, more commonly known today as the Uffizi, was shown only from the outside, as part of a streetscape. The ruins of the Coliseum, which signified Rome and Classical civilization in Grand Tour portraits of Englishmen, did not appear in English zograscope views from the same period. The garden and the exterior of Versailles, which could be balanced by

Figure 1.4   *A View of the Grand Duke's Gallery and Old Palace at Florence, from the River Side* from the *Gentleman's Magazine,*
October 1750. Photo courtesy of The Folger Shakespeare Library.

views of English gardens and palaces, were shown, but the inimitable Hall of Mirrors,
seen later in French zograscope views, never appeared.[20]

The subjects of zograscope views allowed Britain to be on a cultural par with Europe,
with greater London and a few estate gardens standing for modernity and civilization,
while Scotland and Wales were nearby adjuncts providing quaint castles and ruins. John
Bowles and Son cross-listed their four general views of London among the zograscope
views of "European" cities, but not their views of Edinburgh and Glasgow in Scotland.[21]
As for England's provincial cities, the only one advertised as a zograscope view in the
newspaper is Chester, which John Boydell cast as an "antient" [*sic*] walled city on the
Welsh border, with three Welsh castles completing the picture set. Map- and printsellers'

catalogs added Oxford, Cambridge, Portsmouth, Worcester, and Newmarket to the tally of provincial towns, but a total of five provincial towns pales next to the sixty-seven in Samuel and Nathaniel Buck's contemporary series of views, not for use with a zograscope.[22] Even though urban polite society existed outside London, London set the standard, and represented it alone in zograscope views.[23] Zograscope views did not survey all the spaces and places of Britain; they showcased a certain idea of its public space as polite and worthy of its Continental cultural rivals.

The zograscope world circa 1750 excluded references to global trade and exploration. In Bristol, R. Darbyshire sold zograscope views printed in London, but none of these would have shown the important port city that was his home.[24] Industrial areas like Birmingham, Newcastle, and Norwich do not appear, although their mines, mills and machines are subjects of nonzograscope prints at the time. The people who sold maps and views of current events from the War of the Austrian Succession sold zograscope views at the same time, but zograscope views almost never showed war-related sites.[25] Trade, politics, and industry, subjects that were inappropriately businesslike for polite conversation, were also inappropriate for virtual reality viewing in the 1740s and 1750s.

Similarly, sentimental scenes did not appear in English zograscope views. The same cannot be said for Continental zograscope views, which included subjects like prisoners confined in St. Pierre, or episodes from the story of the Prodigal Son. The absence of such subjects in English views must therefore have been due to something other than their suitability for the optics of the zograscope. Joseph Highmore's retelling of Samuel Richardson's *Pamela,* which appeared as a series of prints in 1745, shows that the print medium itself was perfectly suitable to sentimental subjects in England at the time. These prints were not adapted for use with a zograscope, however, and the same holds true for any number of other sentimental prints. Zograscope views were not about sentiment. They were about politeness. English zograscope views depicted those polite places that were suited to neutral public sociability.

In other words, zograscope views exhibited an emotional detachment in their subject matter, and even when they did venture into locations that might seem more picturesque or sensational, those locations tended to be outside England, and outside what became known as the sublime aesthetic. Dramatic, moody prints of England's Peak and Lake Districts, although published in the 1740s and 1750s, were not advertised for use with a zograscope until the late 1760s, well after zograscopes ceased being new media. The curiosity cabinets of eighteenth-century England may have housed exotica, but zograscope views seen by the same people were unashamedly familiar. In contrast, the stereographs of the nineteenth and early twentieth century (see chapter 5) routinely pictured the exotic, hinting suggestively at a later coincidence of stereopsis and imperialist desire.

Zograscope Views and the Mapping of Space

More than being depictions of the public places of polite society described above, zograscope views can be considered mappings of polite public space. Eighteenth-century England saw the urban streets increasingly differentiated by function, with distinct thoroughfares, promenades, shopping streets, and back alleys.[26] Zograscope views helped define the polite and public side of the new urban demarcations, and the polite and public side of their rural counterparts. The ostensible subjects of individual prints are specific public "places," which become generic public "spaces" through the sense of volume and distance brought about by the zograscope. Zograscopes brought this new kind of space indoors, where it could be seen virtually, within the private space of the domestic interior.

"Public" in the context of polite sociability does not mean available to all persons, but available to those men and women considered to be the "right" kind of people.[27] For example, when the *Gentleman's Magazine* first appeared early in 1731, it carried the full title *Gentleman's Magazine or Trader's Monthly Intelligencer*. By 1733, the word *Trader's* had been dropped, not because the contents did not appeal to traders, but because in order to appeal to them, they had to be tacitly subsumed within the category "gentlemen."[28] Similarly, the full-titles of printsellers' catalogs addressed "Gentlemen" as the people who would buy prints to keep for their collections, while the same prints were for "Merchants for Exportation, and Shopkeepers to Sell Again."[29] This does not mean that merchants and shopkeepers did not buy prints to keep, but that when they did so, they played the role of gentleman, not merchant or shopkeeper. These categories could even cross gender lines to a degree. Prints for the genteel handicraft of japanning were specifically designated "for ladies" within a catalog text not because other prints were unsuitable for women, but because trimming and varnishing prints for ornamentation was coded "elite female" in the same way that buying prints in bulk and reselling them at a higher price was coded "middle-class male" (merchant or shopkeeper).[30] By buying (or at least viewing) zograscope views, women and men were participating in an activity implicitly coded "polite society," where no reference could be made to gender or profession in the advertising copy.

Categories that had formerly been perceived as fixed and innate became fluid, inhering in the same person at different times depending on the occasion. For a broadly based commercial society to function, a less hierarchical, less regimented, but still rule-bound way of relating in society had to develop, and in the 1740s and 1750s in England, the category of the "polite" took this role.[31] In order to enter the category of polite individuals in public space, a person had to meet both the monetary admission cost where applicable (e.g., one shilling at Vauxhall Gardens), and the qualitative admission cost by dressing appropriately, knowing how to make suitable light conversation with other polite individuals, and remaining politely detached from his or her surroundings. For literate people with a certain amount of leisure time who wanted a model or guide to polite standards, numerous books and magazines showed the way through words. Zograscope views conveyed standards of politeness in relation to public space through images.

Acts of mapping are acts of visualizing, conceptualizing, recording, representing and creating spaces graphically. It is this living relationship between the observer and the zograscope view that I consider to be "mapping," in that it allows for the active simulation of space, not just its passive representation. The history of *maps* is a technical matter, whereas the history of *mapping* is a matter of examining on the one hand what it is about a culture that determines what a map will look like, and on the other, what the map comes to mean to people in *use*.[32] In the case of zograscope views, the culture in question can be described as a group of private persons coming to think of themselves as part of a single public, the emerging bourgeois public sphere described by Jürgen Habermas and Richard Sennett, for which I am using the contemporary term *polite society*.[33]

Habermas describes culture at the time as something that ceased being part of church and court spheres and instead became commodified for the public who "as readers, listeners, and spectators, could avail themselves via the market of the objects that were subject to discussion."[34] This rational discussion of cultural objects among private individuals as part of a larger public helped shape that public, beginning, in the case of England, in the coffee houses of the late seventeenth century. In Habermas's chronology, the press and professionally written criticism took over the institutional role of coffee houses from about 1750.[35] Also from about 1750, as David Solkin has convincingly shown, painting in and of the public sphere functioned this way, as both a site and an object for discussion.[36] Zograscope views can be understood the same way, as another example of the visual arts playing a role as a commodity available for discussion. Simply restating Solkin's case with a new set of examples misses the unique power of the zograscope, however. Zograscope views could do something visual that other objects and images could not.

In Solkin's mid-century examples, emphasis mostly falls on reading and speaking, only it is images, not printed words, that are being read.[37] Solkin writes that paintings of low and "popular" scenes on display in Vauxhall Gardens were understood by visitors as amusing depictions of what they themselves were not, and paintings on display at the Foundling Hospital were understood as discourses on sympathy. In other words, the pictures were read; the various pictorial elements acted like words in a sentence, where there is necessarily a time delay in grasping the story, since it takes at least a moment or two to figure out what the picture is "about." Focusing on the narrative element like this leaves aside the uniquely visual aspect of images. Namely, there is something visceral and instant about a visual object that can be exemplified by the difference between the meaning of the sequence of words on a page, and the meaning of the layout and typeface of those words. Zograscope views in this light are not commodifications of "culture" in the artistic sense, but commodifications of space, the space in which "culture" exists and proceeds.

As Geographer David Harvey generalizes:

> Social relations are always spatial and exist within a certain produced framework of spatialities. Put another way, social relations are, in all respects, mappings of some sort, be they symbolic, figurative, or material. The organization of social relations demands a mapping so that people know their place. . . . From this it follows that the production of spatial relations . . . is a production of social relations and to alter one is to alter the other.[38]

It was precisely in this sense that the new zograscope views mapped a new spatial relationship, one that accorded with the emerging, constitutive needs of polite society. The relationship between individuals and public space in polite society must be a relationship of spectator and spectacle, not participant and activity.[39] Sites become distanced and neutral sights. Zograscope views provided a model for experiencing space with the same distancing from personal and particular circumstances. Tiny, anonymous people appear over and over in the same kind of vast cosmopolitan spaces.

In order for polite space to come into people's consciousness, it has to first be described, or be made describable.[40] Zograscopes had a super-descriptive quality: The kind of description necessary was qualitative and refered less to the specific place involved than to the type of space that specific place ought to occupy. A particular site might already be known to the observer through having visited or done business there, or through pre-conceived notions from literature; zograscope views worked to counter this knowledge by removing sites from their physical, historical, or literary contexts and allowing them to be regrouped as interchangeable components of a set. Instead of describing a partic-ular place, zograscope views describe a particular kind of space, mapping it out so that a specific lived-in site can be understood as a generic, distanced sight. That is, zograscope views participated in a construction of knowledge about space, knowledge of how public space could be rendered appropriate for polite society.[41] Only a neutral and "empty" kind of space, such as the narrative-free yet apparently real space of the zograscope view, met the requirements of the eighteenth-century discourse of politeness, a point that requires some expansion on the meanings of space and place and their relationship to virtual real-ity and to mapping.

Space can be conceived of in two related ways, one static, and the other dynamic. The difference between the two can be seen in the adjectival form each takes: *spacious* for the static aspect, and *spatial* for the dynamic. Place, in contrast, is always only static and par-ticular. Differences in language and vocabulary complicate the issue. For instance, the French translation of Habermas's *Strukturwandel der Öffentlichkeit* renders the title *L'Espace public* and uses the words *espace* and *sphère* interchangeably in the text, although Haber-mas does not address the issue of abstract, spacious space as such, a kind of space that is central to zograscope views and difficult to articulate in German.[42] Michel de Certeau, for his part, contrasts *espace* and *lieu,* translated as *space* and *place* respectively, in *The Practice of Everyday Life,* yet his concept of abstract space is spatial only, not spacious.[43] Writing in English, David Harvey uses *space* in both senses, but does not recognize their difference.[44] Distinguishing between "spacious" space, "spatial" space, and particu-lar "place" might seem to be little more than arcane word-play, but examining zogra-scope views in terms of these distinctions reveals what is unique about them, and what made them so appropriate to the needs of commercial society operating under a code of politeness circa 1750.

Space as spacious is open and abstract. It carries the illusion of neutrality in its empti-ness and sense of distance. In contrast with this kind of space, a place is particular, en-gaged, sometimes even personified. Place is always full, although its particular contents will vary from perceiver to perceiver. As a fullness, "place" is incompatible with eigh-teenth-century notions of politeness. Politeness requires "delicacy" and an avoidance of all that would seem indecent.[45] It requires spaciousness in which an appropriate distance from particulars can be achieved and maintained. The superficiality of the subject matter and clarity of composition in zograscope views fits this standard, but such polite banality would have been equally apparent in the prints had the zograscope never appeared. Rather, the polite spaciousness that is mapped in zograscope views comes from the erasure of the print surface altogether. The enacted zograscope view with its blurry margins and stereoptic effect is not an image of width-wise expanse, in the manner of a map or panoramic view, but an image of expansion into depth. Thus zograscope views can map space as spacious in ways other pictures cannot, and it is the spaciousness, brought about by the optics of the device in relation to human vision, which aligns the image with "politeness," allowing the zograscope view to participate in the rhetoric of politeness at a deeper level than the subject matter and composition of the print seen with the naked eye.

To illustrate what I mean about politeness through spaciousness, consider the difference between a conventional map of a site and a painting of the same. The conventional map abstracts space into a flat "footprint," making the map a schematic representation of the specific place in question. Space and ease-of-movement on the map can be represented by widening the footprint of streets and canals in relation to the buildings' footprints, but this map remains two-dimensional, without a sense of volume. It is compressed into a flat surface, seen from above.

A painting of the same site, on the other hand, does acknowledge the volumetric quality of space, and represents it through linear perspective. It can only be a representation of space, though, not a simulation or *mapping* of space. Imagine an actual eighteenth-century painting, with light glistening off the varnish, perhaps some feathery touches of pale paint stroked on top as highlights, and visible brushstrokes in areas where the paint is thick. The very fact that it is made of paint means that this picture will always be a surface, and not a space. It bears both the weight of the physical material with which it is made and the immediate physical traces of an individual's hand. Therefore it is too obviously material and too obviously connected with the bravura performance of an artist to be an instance of the trifling, unceremonious, easy detachment that defined politeness.

A similar argument about materiality of surfaces could be made for a print of the same site. The paint is translucent watercolor rather than heavy oils, so now it is the texture of the paper and the graphological quality of the engraving rather than the brushstrokes and pigment that call attention to the two-dimensionality of the representation. This argument holds true only as long as the print is seen with normal vision, though. As soon as it is seen through a zograscope, the picture surface dissolves into three-dimensional space. Instead of the ideal one-eyed observer who must gaze fixedly at a scene, the observer becomes a living two-eyed being who can freely glance at different parts of the scene without destroying the illusion of depth. By erasing the surface of the image, the zograscope both creates spaciousness and allows the observer to relate to it with natural vision, at a polite distance, in a free and easy way.

Unlike spaciousness, the dynamic spatiality of space cannot be conceived of as an empty volume. Michel de Certeau writes of this kind of space that it is "composed of intersections of mobile elements" or "actuated by the ensemble of movements deployed within it." Put in more concrete terms, he says "*space is a practiced place.* Thus the street geometrically defined by urban planning is transformed into a space by walkers" (emphasis in the original).[46] De Certeau's dynamic, living space cannot exist in a single eighteenth-century print seen without a zograscope. Without the zograscope, the image necessarily has the static "being there" of a place in de Certeau's terms. A set of zograscope views, on the other hand, begins to shift the images from place toward space, creating the neither-here-nor-there actuality of "mapping," which de Certeau does not address as such, but which his terminology can help illuminate.

Consider the contrasting pairs de Certeau uses to characterize place versus space[47] as two different ways of relating to the known world:

PLACE	SPACE
seeing	going
map	tour
identified	actualized

In respect to de Certeau's model, zograscope views are an amalgam of place and space, combining characteristics from both sides of what was for him a theoretically absolute divide. With the zograscope, the images are a form of "actualized" "seeing," bringing a type of space to life rather than "identifying" a particular site (place), yet the space can be moved through only up to the limit of the picture's borders. As an arrangement of sites in series, they constitute an itinerary for visually "touring" polite geography (spatial), not a single "map" where sites are simultaneously present, in parallel (place). In their virtual reality mode, zograscope views are not fixed statements that can exist like a "map," apart from the observer (whose visual cortex creates the illusion), but only exist in relation to the person doing the seeing, like a "tour." Yet as a finite set of images, zograscope views provide what de Certeau would consider a maplike sense of place. This allows zograscope views to map space as "polite" according to eighteenth-century understandings of the term: Overbearing sensationalism and overbearing rationalism are both avoided because the relationship with the observer is limited and dynamic at the same time.

Taking the monument in Fish Street as an example again, one can see how walking in the street and zograscopically viewing the street both create space, but only the zograscope space is polite, in the mid-eighteenth-century sense. When actually walking along the street depicted in figure 1.3, the experience that creates the space is tied to the various bodily sensations of feeling feet on paving stones, hearing church bells, smelling the passing horses. The zograscope view of Fish Street, however, limits the actualization of the site to the polite realm. The body's sense-perception remains safely within domestic space—the feel of feet on carpet, the sound of pages turning, the scent of tea—while the eyes and mind can contemplate Fish Street and the Monument as a public, active space.

In addition to the physicality of the body being in Fish Street, the experience of actually walking there is tied to the walker's trajectory. Where you are going, and where you have been is inevitably part of the experience. Thus the "actualized" space when walking down the street includes not only the immediate sight, sound, smell, taste, and feel of being there, but the memory of personal, idiosyncratic sensations from what you have passed (a friend's home, an accident in the roadway . . .) and the anticipation of arriving at your destination. Such a saturated experience is inappropriate to the aesthetic of politeness, a problem rectified in zograscope viewing, where the trajectory is not a seamless series of particular and ever-changing places transformed into spatial space by walking. It is a series of discrete prints that can be viewed in any order, with the same generic perspective taken up by the observer again and again. The view of the Monument might be seen as one of a group of thoroughfares, or a group of landmarks, or just a random grouping of prints in the order in which they happen to have been put back in their box. A zograscope image is not a deep understanding of a particular place, but one of many interchangeable components in a system that can collectively be thought of as polite public space. Space itself is being commodified.

Nor are mid-eighteenth-century English zograscope views narratives in the common meaning of the term. A mid-eighteenth-century view of a church interior, for instance, would never show a service in progress, nor would it draw attention to the actions of particular figures. The people in such prints tend to be small, positioned in groups of two or three, chatting with each other, gazing at the architecture, or sometimes unobtrusively giving and receiving alms. Views of palaces do not feature members of the royal family but rather show anonymous figures from a distance, going about their business or enjoying the vistas. Catholic pageantry is absent from St. Peter's in Rome. The Lord Mayor's coach might appear outside the Mansion House in London, but his presence is neither visible in the print nor mentioned in its caption. There is no obvious story or moral to be had in the print itself, just a view of a public site in an objective, impersonal, and therefore polite mode. The only story told is one of space.

Zograscope views take many different places—a Scottish castle, a Roman piazza, a London church, a Paris bridge—and cast them all into the same kind of generic space. The prints' captions give little information other than the name of the city and the principal buildings depicted. The point of the captions is to familiarize the sites in a general way, without spelling out political and historical details like the captions of Bucks' views or describing actual events in the scene, which could create a narrative with overt appeal to the depths of heart or mind. Observers actualize this space through the natural eye movements of normal distance-vision while manipulating the zograscope and prints on the table in front of them. The spaces depicted are thus rendered polite, uniform, real, and manageable. The new kind of space is distanced from the observer, yet the observer is still implicated thanks to his or her involvement in arranging the prints and realizing the simulation. Public space is rendered palatable for domestic consumption.

A month after the fact, Peter Brookes and Robert Sayer advertised their zograscope view of the Green Park fireworks in celebration of the Treaty of Aix-la-Chapelle as something that would provide "a just Idea of the Structure, and its Situation, as well as the Performance itself."[48] Given that the "performance itself" had ended prematurely when a charge misfired, engulfing the structure in flames and sending the carefully orchestrated light show to a chaotic and sudden conclusion, the last part of the advertisement reads more like a disincentive to buy. However, a "just idea" of the performance need not be a retelling of the narrative of events (although I do not wish to suggest that Brookes and Sayer actually thought this through when placing the advertisement in the newspaper). In the same way that a portrait photo taken from an unsympathetic angle in unflattering light may not "do justice" to the sitter, while another (even if highly retouched) will, the justice of the zograscope view of the fireworks is that it shows public space as it ought to be perceived. That is, a zograscope view of the fireworks shows a public display in the familiarity of a private, domestic setting, where it can be simultaneously made real and placed at a polite distance.[49]

Little evidence about actual viewing practices from the 1740s and 1750s survives, but because the lens lets only one pair of eyes at a time see the illusion, and because the zograscope was simple to set up, we can infer that they were often viewed alone, or in small groups. Lady Luxborough writes of "an optical glass which I have lately purchased" as something where she herself places the prints in the machine, and she makes no mention of other people being with her.[50] Although zograscope views provided a spectacular vision of public space, they were something to be experienced in a private setting, either alone or in small groups. Like the novel, magazine, and newspaper, zograscope views brought public spaces to individuals in a private setting. Also like the novel, magazine, and newspaper, the zograscope did not attempt a total isolation of the individual from his or her surroundings. Although important for the illusion of depth in the picture, "immersion" in the virtual reality sense was not enhanced to the degree it could have been with the zograscope. Bordering the prints in black did isolate the illusory scene within the field of view when looking through the lens of the zograscope, but common stand-type and book-type zograscopes did not seal the print away from the rest of the room. Drawing-room observers manipulated the prints themselves. They may even have been the ones who added color to the views, using paints sold ready-prepared for coloring prints at home. There was no hiding the constructedness of the illusion.

Observers claimed that seeing a zograscope view was like having "the Life itself be there,"[51] but it should be emphasized that it was like having the life itself be there under controlled circumstances, set off for contemplation by the wonders of the modern scientific instrument. A simulation not only creates an effect, it also isolates that effect and brings it forth for examination.[52] It is for this reason that the zograscope view and zograscope viewing had to maintain a link with the domestic interior. The zograscope produced specimens of virtually real space that exemplified polite detachment from personal particulars yet did not require arcane aesthetic knowledge to understand. They could be talked about naturally and easily in terms of optical effects (zograscopes figure prominently in both the catoptrics and dioptrics sections of popular science texts at the time), in terms of the sites depicted, and in terms of the group of sites as a whole.

It is in this last respect, as items making up a visual collection, that the final piece of the puzzle allowing zograscope views to generate an understanding of polite public space falls into place. By compiling and dividing into individual views a theoretically finite number of polite public spaces in Western Europe, each in the same horizontal format on a half-sheet of paper, a set of zograscope views was a collection of like things. Devoid of the sentimental properties of a souvenir, except in rare cases where the observer did happen to have purchased the print when visiting the site depicted, they could be arranged and rearranged to create something new.[53] Zograscope views were examples of the kind of space one ought to know about, divided up into equivalent units for ease of understanding and integrated by the observer into a whole. Unlike souvenirs, where the gaps between the objects that make up the "story" are filled in by the individual's particular memories, thoughts, and desires, the emptiness of the gaps between zograscope views were integral to the experience. They encouraged the observer to perceive the prints with the discreteness and detachment of politeness instead of as a seamless narrative.

Conclusion

In theory, constructing knowledge of space by emptying places of previous associations, then filling the resulting empty space with new values, will create a new place.[54] In practice, polite space would always exhibit a tension between abstract ideals and actual human impulses.[55] A writer in the *World* (19 December 1754) pointed out with tongue-in-cheek pragmatism, "It is not virtue that constitutes the politeness of a nation, but the art of reducing vice to a system that does not shock society." Specifically, certain aspects of the world had to be concealed in order for polite society to function. Zograscope views provided a model for this concealment by supplying polite society with new mappings of space. They masked the individually specific experiences each person had of a place in order to provide the illusion that there was one proper, detached, way of experiencing space. They showed what kind of identity (of the many any one person has) was appropriate in public: one that denied the immediacy and physicality of the body, the actuality of sore feet while walking, of having to watch for traffic when crossing the street, of losing one's way. Conventional maps, which do a good job of showing a territory as a single bounded entity, would be of little use to a community that does not exist within territorial boundaries. Polite society needed a different kind of map, one that set out and unified the *kind* of space involved. In accord with the needs of modern society, this space appeared as an absolute structure that individuals could (ideally) understand in the same way.[56] More insidiously, perhaps, zograscope views naturalized the idea that the domestic world (the space of the viewing) and the public world (the space of the view) were related, but separate. In order for the political public sphere described by Habermas to develop out of the bourgeois public sphere, it was necessary that the separation be absolute and gendered, a change of mindset that the zograscope facilitated.

Zograscopes are sometimes portrayed as precursors of the cinema and television, in that they presented an illusionistic image of one space to observers in another space.[57] In a technological sense, this is undoubtedly true. It is also true of many later zograscope images. However, unlike conventional cinema and television, nothing obviously *happens* in English zograscope views circa 1750. The three-dimensional illusion of space and the limited subject matter, not an internal narrative, create the experience. In this experience, certain spaces are real-but-not-real, close at hand but at a distance, lifelike but without the personal identification and emotion that bring plays and novels, as well as cinema and television, to life.

Considering early zograscopes as part of cinema history also presumes that the zograscope view continued to develop along the same lines until eventually being replaced by a newer and better virtual reality device. In England, at least, this was not the case. After the 1750s, zograscope views suddenly disappeared from newspaper advertisements. The *Universal Magazine* made no more mention of them. The *Gentleman's Magazine* stopped promoting them. The *Copper Plate Magazine* (which issued numerous prints whose titles began "Perspective View of . . ." between 1774 and 1778) ignored them completely. When the *Gentleman's Magazine* for August 1765 included a view of Vauxhall Gardens almost identical to one published as a full-sized zograscope view in 1751 or 1752, the magazine's title page called it "a beautiful Perspective view" of the garden, "elegantly engraved on Copper."[58] Had this plate appeared in the magazine fifteen years earlier, its suitability "for the Concave Mirror or optical Machine," the formula given for perspective views in title pages from 1749 through 1751, would have been highlighted, not its beauty and elegance. Many people who owned zograscopes apparently continued to color the later *Gentleman's Magazine* plates for the devices on their own initiative,[59] but the magazine had changed its editorial mien (always geared toward people who wished to be considered Gentlefolk) toward fine art aesthetics when publishing view prints.

Fashions in any age only last so long, and five or ten years of attention for the same kind of images in newspapers and magazines is quite a feat at a time of increasingly rapid change. The taste for "sentiment" that in some sense succeeded "politeness" happened to be incompatible with zograscope viewing, and although some publishers added sentimental genre scenes and landscapes to their catalogs of zograscope prints, the interposition of a viewing device between observer and image opposed the intimacy demanded by connoisseurs of sentiment. The needs of polite society had changed. Publisher John Boydell turned away from zograscope views completely after mid-century, intending instead to elevate the tone of English print culture with "Designs in Genteel Life" in emulation of French prints.[60] Ironically, he had to turn from a native English genre to a French one in order to promote national pride, thanks to changing ideas about what counted as cultural achievement. Britain remained known for precision scientific instruments, but the link between optical instruments and images no longer had a cultural cachet.[61]

Although Robert Wilkinson carried much of John Bowles's stock of zograscope views on into the nineteenth century, often publishing them jointly with the firm of Bowles and Carver (successors to Carington Bowles, who had succeeded Thomas Bowles), Bowles and Carver's 1795 catalog indicates that they added very few prints to their zograscope repertoire. Robert Laurie and James Whittle, successors to Robert Sayer, and in turn their successor, Richard Holmes Laurie (Robert's son), did add some new zograscope views, notably sentimental genre scenes, specific episodes from the Napoleonic Wars, and classical landscapes. Five zograscope prints under the heading "Landscapes" appeared in Laurie and Whittle's 1795 catalog: "The Travellers," "The Fowlers," "A Conversation," "Rural Pleasure," and "Wreeken Hill, Shropshire."[62] These five appeared as a group again in the 1824 catalog, along with two more: "Plundering Country Pheasants" and "Summer, Figures with a Cascade from Watteau."[63] Other new zograscope scenes in 1824 included "The Italian Fisherman," "The Judgement of Midas," "Returning from Market," "The Journey," and "The Repose." And although the generic scenes of Paris engraved before 1750 reappeared, now they were joined by prints like "The entrance of his Majesty Louis XVIII into Paris, 3d May, 1814." No longer mappings of polite public space, these new zograscope views (with the exception of "Wreeken Hill, Shropshire") had narrative, not space, at their core.

Other zograscope prints departed from the mapping of space in a different direction, favoring high art landscape (but not high art virtuoso engraving technique) rather than narrative genre. Some time after 1824, Laurie added picturesque views of the Lake District to his catalog of zograscope prints, plus a group of "Twelve very Pleasing Landscapes" after paintings from the seventeenth and eighteenth centuries. Prints after these paintings would have been available for use in 1750, but their picturesque subject matter made them inconceivable as zograscope views then.

New forms of virtual reality experiences arose toward the end of the eighteenth century, ones with very different relationships to public and private space than zograscopes had. In 1787, Robert Barker patented what soon became known as the "panorama," a life-size wrap-around painted scene with the perspective calculated to make paying visitors feel they were seeing the real thing.[64] Such spectacles served as a form of public entertainment, something vastly different from the private, protected viewing space of the zograscope. Likewise, from 1820 the Cosmorama—a permanent exhibition room with individual oil paintings exhibited behind a series of zograscope lenses set into a wall—became a public destination.[65] The pictures spanned a range of subjects, related only by the observer's movement from one to the next. Special lighting effects added fire, gloom, or sunshine to the scenes. This emphasis on variety, sensationalism, and physical movement contrasted with mid-eighteenth-century zograscopes, where the experience involved seeing the same kinds of polite spaces over and over while seated or standing in someone's home. The optical principles might have been the same. The experience of space ws not.

Panoramas, Cosmoramas, Dioramas, Physioramas . . . the list of public spectacles whose illusion consisted of imagining oneself to be looking into real space could go on right up to the cinema, bringing in small-scale nineteenth-century persistence of vision devices like the zoetrope, phenakistiscope, and praxinoscope along the way. The point here, though, has not been to place zograscope views into a seamless chronology of new media. It has been to describe the very particular role of zograscopes in England in the years around 1750. When stereoscopes came to public attention in the middle of the nineteenth century (see chapter 5) they encompassed a new vision, one that included empire and world trade.

Notes

1. "Thut man aber in das Sehe-Loch ein ander *vitrum convexum,* so etwan einen solchen *radium* hat, als das Loch von dem *opposito albo* stehet, so sieht man die *objecta,* also vorher gedacht, wie sie von *aussen libero oculo* vorkommen, in der Weite, Breite, Kündung, und Distantz. Ist ein neues *inventum,* welches auch in der *obscura camera* zu gebrauchen," Johann Christoph Kohlhans, *Neu-erfundene mathematische und optische Curiositäten* (Leipzig, 1677), 295. William Molyneux, *Dioptrica Nova* (London, 1692). This chapter is based on my doctoral dissertation, *Zograscopes, Perspective Prints, and the Mapping of Polite Space in Mid-Eighteenth-Century England,* Stanford University, Department of Art and Art History, April 2000.

2. Most zograscope prints measure approximately 11 × 17 inches, although those published by the *Gentleman's Magazine* were considerably smaller. Contemporaries agreed that the prints had to be "properly coloured" for viewing with the device. In practice, this meant outdoor scenes with a predominance of blue, green, and brownish-yellow, with a bright blue sky fading to pink, and touches of red and bright yellow here and there. Indoor scenes limited intense color to drapery and a few architectural details, with only a pale wash elsewhere.

3. Zograscopes were also used in other countries in the eighteenth and nineteenth centuries, notably Continental Europe, the United States and Japan. This chapter addresses the particular story of polite society in England circa 1750, and one of my assumptions is that zograscopes possessed different implications for different viewers, in other times, and other places.

4. See Shojiro Nagata, "How to reinforce perception of depth in single two-dimensional pictures," in *Pictorial Communication in Virtual and Real Environments,* ed. Stephen R. Ellis, Mary Kaiser,

and Arthur J. Grunwald (London and Washington: Taylor and Francis, 1993), 527–545; Harold Schlosberg, "Stereoscopic Depth from Single Pictures," *American Journal of Psychology* 54, no. 4 (October 1941): 601–605; Alfred Schwartz, "Stereoscopic Perception with Single Pictures," *Optical Spectra*, September 1971, 25–27.

5. William Emerson, *Perspective: or, the Art of drawing the Representations of all Objects upon a Plane* (London: 1768), iv–v; Henrietta Knight, *Letters Written by The Late Right Honourable Lady Luxborough, to William Shenstone, Esq.* (London, 1775), 18 [in a letter dated 1748].

6. Jonathan Crary, *Techniques of the Observer: On Vision and Modernity in the Nineteenth Century* (Cambridge, Mass.: MIT Press, 1990). In order to fit the zograscope in with the type of evidence Crary uses, it should be added that although the zograscope was known at the time to produce a virtual image at least in part through binocular stereoscopy, the essential nature of this image was not a subject for discussion, and when stereoscopes eventually did come about, they were never described as perfected zograscopes. Even David Brewster's 1856 history of the stereoscope, which goes back to Euclid on binocular vision, does not mention the zograscope.

7. David H. Solkin, *Painting for Money: The Visual Arts and the Public Sphere in Eighteenth-Century England* (New Haven: Yale University Press, 1993); Dena Goodman, *The Republic of Letters: A Cultural History of the French Enlightenment* (Ithaca: Cornell University Press, 1994); Ann Bermingham and John Brewer, eds. *The Consumption of Culture, 1600–1800* (New York: Routledge, 1995).

8. Jürgen Habermas, *The Structural Transformation of the Public Sphere: An Inquiry into a Category of Bourgeois Society,* ed. Thomas McCarthy, trans. Thomas Burger with assistance of Frederick Lawrence (Cambridge: MIT Press, 1989).

9. For "politeness" as a way of life, see Paul Langford, *A Polite and Commercial People: England 1727–1783,* ed. J. M. Roberts (Oxford: Clarendon Press, 1989), and John Brewer, *The Pleasures of the Imagination: English Culture in the Eighteenth Century* (New York: Farrar, Straus, Giroux, 1997).

10. Josef W. Konvitz, "The nation-state, Paris and cartography in eighteenth- and nineteenth-century France," *Journal of Historical Geography* 16, no. 1 (1990): 3–16; Miles Ogborn, *Spaces of Modernity: London's Geographies, 1670–1780* (New York and London: The Guilford Press, 1998).

11. *Public Advertiser,* 27 February 1754.

12. *St. James's Evening Post,* 2–4 April 1747; *General Advertiser,* 19 January 1748.

13. The earlier date proposed by C. J. Kaldenbach, "Perspective Views," *Print Quarterly* 2 (1985): 86–104, comes from a misunderstanding of the term "perspective view."

14. *Two Hundred and Six Perspective Views Adapted to the Diagonal Mirrour, or Optical Pillar Machine.* London, 1753.

15. John Stow, *A Survey of the Cities of London and Westminster, and the Borough of Southwark,* 6th ed., 2 vols. (London, 1754, 1755).

16. *London Evening–Post,* 22–24 January 1751.

17. Timothy Clayton, *The English Print, 1688–1802* (New Haven and London: Yale University Press for The Paul Mellon Centre for Studies in British Art, 1997), 157–161.

18. Benjamin Martin, *A New and Compendious System of Optics, in Three Parts* (London, 1740), 290–291.

19. From a study of the British Library's Burney Collection. This is the number of titles, not the number of advertisements; only the first occurrence of each advertisement was used to tally the totals. Small-sized zograscope views are not included.

20. For instance, "Galerie des Glaces," illustrated in *Les Vues d'optique,* exhibition catalog, Chalon-sur-Saône, Musée Nicéphore Niépce, 10 December 1993–March 1994.

21. John Bowles and Son, *Catalogue of Maps, Prints, Copy-Books, &c.* (London, 1753).

22. The Bucks' "Principal Series of Town Prospects," executed between 1728 and 1753, consists of 83 sheets: 73 of provincial cities (6 are shown twice), 8 of London and environs, and 2 of St. Michael's Mount, Cornwall. See Ralph Hyde, *A Prospect of Britain: The Town Panoramas of Samuel and Nathaniel Buck* (London: Pavilion, 1994).

23. Rosemary Sweet, *The Writing of Urban Histories in Eighteenth-Century England* (Oxford: Clarendon, 1997).

24. *General Advertiser,* 20 February 1749.

25. The exception is a set of six prints entitled *The Manner of besieging a Town,* which was sometimes advertised for use with a zograscope, sometimes not. Thomas and John Bowles do not identify the town in their advertisements, but Robert Sayer says it is Barcelona.

26. Penelope J. Corfield, "Walking the City Streets: The Urban Odyssey in Eighteenth-Century England," *Journal of Urban History* 16, no. 2 (February 1990): 148.

27. In the sense that I use the words *public* and *private* in reference to place and space, I do not mean to evoke specific rights and responsibilities of citizens or the administrative state. I also do not mean interiority versus relations with other people. Public and private here simply refer to different arenas of sociability for polite society. Nondomestic social spaces like streets, parks, and churches are considered "public." Domestic social spaces such as the drawing room and the library are considered "private."

28. Langford, *A Polite and Commercial People,* 65; Alison Adburgham, *Women in Print: Writing Women and Women's magazines From the Restoration to the Accession of Victoria* (London: George Allen and Unwin, 1972), 79.

29. Robert Sayer, *Robert Sayer's New and Enlarged Catalogue* (London, 1766).

30. For example, John Bowles, *Catalogue of Maps, Prints, Copy-Books, &c.* (1768), 166.

31. Langford, *A Polite and Commercial People,* and Tom Wiliamson, *Polite Landscapes: Gardens and Society in Eighteenth-Century England* (Baltimore: The Johns Hopkins University Press, 1995).

32. Denis Cosgrove, "Introduction: Mapping Meaning," in *Mappings,* ed. Denis Cosgrove, Critical Views (London: Reaktion Books, 1999), 1, 9, 17.

33. Habermas, *Structural Transformation of the Public Sphere;* Richard Sennett, *The Fall of Public Man* (New York: Norton, 1976); Langford, *A Polite and Commercial People.*

34. Habermas, *Structural Transformation of the Public Sphere,* 36–37.

35. Ibid., 51.

36. Solkin, *Painting for Money,* especially chapters 3 through 5.

37. For a critique of Habermas's emphasis on readers and speakers rather than, for instance, performers or spectators, see 117–137; and Joan B. Landes, "The Public and the Private Sphere: A Feminist Reconsideration," in *Feminists Read Habermas,* ed. Johanna Meehan (New York: Routledge, 1995), 100–101.

38. David Harvey, *Justice, Nature and the Geography of Difference* (Oxford: Blackwell, 1996), 112.

39. Ogborn, *Spaces of Modernity,* 109.

40. Walter D. Mignolo, *The Darker Side of the Renaissance: Literacy, Territoriality, and Colonization* (Ann Arbor: The University of Michigan Press, 1995), 227.

41. The argument for maps as knowledge construction as opposed to knowledge transfer comes from Alan MacEachren, *How Maps Work: Representation, Visualization, and Design* (New York: Guilford, 1995), as cited in Jeremy Black, *Maps and Politics* (Chicago: University of Chicago Press, 1997).

42. Jürgen Habermas, *L'espace public: archéologie de la publicité comme dimension constitutive de la société bourgeoise,* ed. Miguel Abensour, trans. Marc B. de Launay (Paris: Payot, 1986).

43. Cf. *"spatial"* and *"spacieux"* in *Le Grand Robert,* and "spatial" and "spacious" in the *Oxford English Dictionary. Spacieux* no longer has the indefinite, abstract meaning it once did in French, and still does in English. ("Spacieux ne se dit plus guère d'un espace libre, de la mer . . ., du monde." [*sic* ellipsis], *Le Grand Robert,* 2d ed., s.v. "spacieux").

44. Harvey, *Justice, Nature and the Geography of Difference.*

45. "Essentials of Politeness or Good-breeding," *Scots Magazine* 3 (December 1741): 554–556.

46. Michel de Certeau, *The Practice of Everyday Life,* trans. Steven Rendall (Berkeley: University of California Press, 1984), 117.

47. Ibid., 118ff.

48. *The General Advertiser,* 27 May 1749.

49. For the difference between public and private spectacle, see Barbara Maria Stafford, "'Fantastic' Images: From Unenlightening to Enlightening 'Appearances' Meant to Be Seen in the Dark," in *Aesthetic Illusion: Theoretical and Historical Approaches,* ed. Frederick Burwick and Walter Pape (Berlin and New York: Walter de Gruyter, 1990), 176.

50. Knight, *Letters . . . by Lady Luxborough,* 18.

51. Martin, *New and Compendious System of Optic,* 290.

52. Gilles Deleuze, "The Simulacrum and Ancient Philosophy," in *The Logic of Sense,* ed. Constantin V. Boundas (New York: Columbia University Press, 1990), 263; first published 1967.

53. Susan Stewart, *On Longing: Narratives of the Miniature, the Gigantic, the Souvenir, the Collection* (Durham: Duke University Press, 1993), 152.

54. Yi-Fu Tuan, *Space and Place: The Perspective of Experience* (Minneapolis: University of Minnesota Press, 1977), 6.

55. John Brewer, "'The most polite age and the most vicious': Attitudes towards culture as a commodity, 1660–1800," in *The Consumption of Culture, 1600–1800,* ed. Ann Bermingham and John Brewer (London: Routledge, 1995), 341.

56. For a description of modern Western capitalist space, see Harvey, *Justice, Nature and the Geography of Difference,* 224.

57. For example, Olive Cook, *Movement in Two Dimensions* (London: Hutchinson, 1963); C. W. Ceram, *Archaeology of the Cinema,* trans. Richard Winston (London: Thames and Hudson, 1965); John Barnes, *Barnes Museum of Cinematography, Catalogue of the Collection, Part One, Precursors of the Cinema* (Saint Ives, Cornwall: Barnes Museum of Cinematography, 1967); Paul Genard, *Cinema d'où viens-tu? Évolution d'une technique . . . naissance d'un art* (Lyon: Centre regional de recherche et de documentation pédagogiques de Lyon, 1975); Maria Adriana Prolo and Luigi Carluccio, *Il Museo Nazionale del Cinema, Torino* (Turin: Cassa di Risparmio di Torino, 1978); *Prima del cinema: Le Lanterne Magiche, La collezione Minici Zotti,* exhibition catalog, Venice, Pedrocchi, 26 March– 26 June 1988; Carlo Alberto Zotti Minici, *Il Mondo Nuovo: Le meraviglie della visione dal '700 alla nascita del cinema* (Milan: Mazzotta, 1988); Hermann Hecht, *Pre-Cinema History: An Encyclopaedia and Annotated Bibliography of the Moving Image Before 1896* (London: Bouwker Saur in association with the British Film Institute, 1993); *Les Vues d'optique* (Chalon-sur-Saône: Musée Nicéphore Niépce, 1993).

58. *Gentleman's Magazine* 35 (August 1765).

59. *A Complete List of the Plates and Wood-Cuts in the Gentleman's Magazine* (London, 1821) says of the Vauxhall plate "many . . . were taken from the Magazines to be put into Optical Machines; and are therefore become scarce" (vii).

60. *Public Advertiser,* Numb. 6342. Tuesday, February 25, 1755.

61. See Barbara Maria Stafford, *Artful Science: enlightenment entertainment and the eclipse of visual education* (Cambridge: MIT Press, 1994).

62. *Laurie and Whittle's Catalogue of New and Interesting Prints* . . . (London: Robert Laurie and James Whittle, 1795).

63. Richard Holmes Laurie, *Catalogue of Perspective Views, Coloured for the Shew Glass, or Diagonal Mirror* (London: R. H. Laurie, 1824).

64. Ralph Hyde, *Panoramania: The Art and Entertainment of the "All-Embracing" View,* exhibition catalog, London, Barbican Art Gallery, 3 November 1988–15 January 1989, 45.

65. Richard Daniel Altick, *The Shows of London* (Cambridge: Belknap, 1978), 211–214.

2 Heads of State: Profiles and Politics in Jeffersonian America

Wendy Bellion

Americans awoke on the morning of November 8th, 2000, to the startling news that the presidential election of the previous day had failed to produce a clear winner. The situation was blamed in part upon the mechanical instruments and methods that voters had used to signal their choice for president. Over the following months, the television and print media swirled with images of the suspected villains: faulty lever-operated machines, punchcard ballots, butterfly ballots, hanging chads, recounts, undervotes. Disenfranchised and disillusioned, angry citizens called for new voting techniques and counting tools to ensure correct political representation in the future.

Nearly two centuries earlier, Americans equally anxious about accurate representation seized upon a new mechanical invention—the *physiognotrace*—as the answer to their concerns. The physiognotrace was a drawing machine, not an electoral instrument, but the era's deepest political worries were etched onto its simple metal frame. Literally used, as the name suggests, to trace an individual's physiognomy, the physiognotrace produced four identical, miniature silhouettes or *profiles.* The device and the images it yielded were praised in the descriptive terms of *actual representation,* a period rhetoric that optimistically imagined political representation to be direct, particular, and true. Although it was not a voting machine, the physiognotrace, as this essay argues, equipped any willing citizen to enact a fantasy of Jeffersonian political subjectivity.

Peale's physiognotrace

1802 was a fortuitous year for Charles Willson Peale, portraitist and proprietor of the Philadelphia Museum. In March, Peale won permission from the state legislature to move his massive *Wunderkammer* of art and natural science from the crowded rooms of Philosophical Hall, headquarters of the learned American Philosophical Society, to the adjacent Pennsylvania State House, national icon of political independence.[1] The formidable task of transferring the museum was largely completed by the year's end, and as he set about rearranging his collections, Peale took the occasion to announce his latest acquisition. "Friendship esteems as valuable even the most distant likeness of a friend," began his December 28th advertisement in the newspaper *Aurora:*

> The ingenius Mr. John I. Hawkins, has presented to C. W. Peale's Museum, an invention of a *physiognotrace,* of so simple a construction, that any person without the aid of another, can in less than a minute take their own likenesses in profile. This curious machine, perhaps, gives the truest outlines of any heretofore invented, and is placed in the Museum for the visitors who may desire to take the likenesses of themselves or friends.[2]

Two weeks later, Peale sat a plaster bust of his own distant friend, Thomas Jefferson, before the "curious machine" and traced the simulacrum of the president's physiognomy. He mailed the results to Jefferson along with a letter that stressed the mechanical utility of the invention and the social utility of its inventor, the English émigré John Isaac Hawkins. "His ingenious Mechanical powers will be of great advantage to America if we can keep him," Peale enthused, noting that Hawkins intended to return to England in the spring. "His republicanism is our only securiety [*sic*] for possessing of him."[3] Jefferson, intrigued, requested further details about the instrument ("I presume no secret is made of it as it is placed in the museum"), and on January 28, 1803, Peale wrote again with a sketch and description of Hawkins's device (figure 2.1).[4] He explained that the physiognotrace was "made strong" in anticipation of the numbers of people that would be handling it. The machine consisted of a pantograph mounted on a vertical wooden board that could be adjusted to an individual's height. The sitter settled himself into a chair, steadied his head against a small concave support, and inserted his shoulder into a semicircle cut into the base of the board. He then grasped a gnomon that Peale termed the "Index"—a five-inch length of brass attached at a perpendicular to the pantograph's lowest joint—and, in an unbroken linear sweep, surveyed his own features. As the index skimmed from chin to forehead, the pantograph translated its movements to a steel stylus that incised a reduced outline of the head, only several inches high, upon a twice-folded piece of white paper affixed to the top of the board. Next, an attendant carefully removed the paper from the board, cut along the outline and removed the interior shape, and then unfolded the paper to reveal four identical hollow heads. Each section was detached and placed against a background of dark paper or cloth to create the flat black-and-white images that Peale and his contemporaries called *profiles* (figure 2.2).

Figure 2.1   Charles Willson Peale, Explanation of Mr. Jno. I. Hawkins Physiognotrace, 1803. Courtesy of the Thomas Jefferson Papers, Manuscript Division, Library of Congress.

The physiognotrace was a boon for Peale. As the art historian David Brigham has observed, the machine brought new visitors and their money to the Philadelphia Museum. Use of the physiognotrace was free to all who paid an entrance fee of twenty-five cents, although Peale charged one cent to cover the expense of the paper. For an additional six cents (later raised to eight), visitors could employ Moses Williams (figure 2.3), Peale's manumitted slave, to trace and cut their profiles.[5] In April 1803, just four months after installing the device, Peale boasted that "much company continue to visit the Museum to get their profiles taken—In fact, it has been the best article to draw company, that I ever had, the Idea of getting a likeness at the cost of only *one Cent,* has a happy effect; such is the love we have of our pretty faces."[6]

Figure 2.2   Silhouette of Rubens Peale. Courtesy of the Library Company of Philadelphia.

Moses Williams, cutter of profiles

Figure 2.3   Silhouette of Moses Williams. Courtesy of the Library Company of Philadelphia.

The physiognotrace also stirred excitement well beyond the walls of Peale's museum. Hawkins patented his invention and extended permission for its use to a handful of individuals who journeyed afield to Charleston, New York, and Boston. But in violation of Hawkins's patent and much to Peale's frustration, entrepreneurs along the eastern seaboard quickly recognized the instrument's lucrative potential and began toting home-made physiognotraces on their itinerant ramblings for work. Profiles were a fast and easy way to earn additional income for painters such as Edward Malbone, who had labored to find clients willing to commission a miniature portrait.[7] Young Ethan Allen Greenwood, who set up his machine in New England taverns, was among some thirty profile portraitists active in that region by 1810.[8] Further south, Peale's son Raphaelle claimed to have produced several hundred thousand profiles with the "facie-trace" he brought to the busy port cities between Baltimore and Savannah.[9] As early as June 1803, Charles Willson Peale could speak without exaggeration of a "rage for profiles"; two years later, he remained in awe of the instrument's phenomenal popularity. "The Phisiognotrace has done wonders," he reported to Hawkins. "Profiles are seen in nearly every house in the United States of America, never did any invention of making the likeness of men, meet so general approbation as this has done."[10]

Public enthusiasm for the physiognotrace turned in part upon the machine's ease, speed, and low expense. The instrument was democratic, as scholars have commented, in the sense that it was affordable and available to a social spectrum of Americans who had neither the means nor use for an expensive portrait in oil paints. However, frequent comments about the "truth" and immediacy of profile likenesses also suggest a profound interest in the instrument's representational properties.[11] Peale consistently described profiles as "perfectly correct" resemblances and attributed this quality to the procedures of self-operation and mechanical circumscription.[12] His early references to the physiognotrace as an "index" are especially telling, for the word implied an empirical capacity to measure, register, and locate an individual. Peale's sons and peers endorsed the physiognotrace in similar terms. Much as Jefferson had supposed that the principles of the device were evident to anyone interested, Raphaelle Peale and his brother Rembrandt promised "to shew the *modus operandi*" of a physiognotrace that they exhibited alongside their father's great paleontological find, the skeleton of a mastodon, in Baltimore during 1804.[13] Elsewhere, Raphaelle swore—perhaps with a hint of speciousness, for his own mimetic talents were considerable—that the device drew profiles "more accurate than can be executed by the hand of the most eminent artist in existence."[14]

This rhetorical insistence on direct and correct depiction configured the physiognotrace as an instrument of actual representation. Indeed, as this essay posits, there is reason to suppose that during the first few years of the nineteenth century, ideas about visual and political representation synchronized around the physiognotrace and the spartan form of the profile portrait.[15] A glance at the conceptual structures of artistic and political representation helps illuminate the problem at hand. The literary critic Christopher Prendergast notes that representation has historically been understood to function in two ways: through *simulation,* in which something absent is spatially or temporally "represented" or "made present again" by way of imitation, resemblance, or illusion; and through *substitution,* in which one agent "stands [in] for" another. Prendergast aligns artistic representation with simulation and political representation with substitution, but observes that both modes of representation can "oscillate" between the two meanings.[16] Indeed simulacra, as semiotic critics have long observed, are but one form by which art represents: instead of or in addition to operating iconically, visual signs may also function as arbitrary *symbols,* so that, for example, a red traffic light means "stop" by cultural con-

vention; or they function as *indexes,* in which the sign bears a physical mark of or phenomenal connection to its own production, such as an artist's fingerprint on the surface
of a clay sculpture.[17] Peale and his contemporaries saw both iconic and indexical properties in profiles. As a "likeness," the profile was understood to resemble an individual's appearance. At the same time, a powerful component of the profile's efficacy was thought
to reside within its status as a material document of an act of representation—the sitter's
self-delineation at the physiognotrace—that was flush with notions of adjacency and accuracy. The following pages elaborate upon these points.

Just as the idea of artistic representation requires some dissection, political representation may be defined as substitution only in the most general terms. A representative
may stand in for one person, as a lawyer arguing a case in a courtroom; or for a group of
people, as a legislator stands for constituents in Congress; or for an abstract entity, such
as property or a state.[18] Moreover, the practical means by which a representative acts as
a legal substitute is partly contingent, as scholars of the Revolutionary era have shown,
upon historical or cultural circumstances. In 1776, newly declared Americans protested
that their interests were not even "virtually" represented by the British crown; in 1787,
advocates of a united federal body encouraged governance by a "natural aristocracy" capable of steering the republic above faction and self-interest; by 1800, Jefferson's supporters were leading a charge for democracy. In the political culture of Jeffersonian
Republicanism, representation was idealized as local, transparent, and direct. Unlike the
British model of virtuality, which held that representatives were beholden not to individuals but rather to shared national interests, proponents of actual representation envisioned delegates as agents of the citizenry and subjected them to close monitoring.
Virtual homogeneity, in short, was no longer any match for the proof of perceptible ties
and similarities.[19]

Although "the artistic and political senses [of representation] must be distinguished,"
Prendergast writes that their conceptual structures may occasionally meet.[20] W. J. T.
Mitchell suggests that theatrical role playing, which involves degrees of both simulation
and substitution, is one forum in which these two modes of representation can come together.[21] The physiognotrace was another. The ways in which this device implicated artistic and political representation can be visualized as a Venn diagram in which the two
spheres of representation, in the historical parameters briefly outlined above, overlap in
the areas of similitude and indexicality. In this field of intersection, where representational truth meant keen resemblance and resemblance was qualified as proximity, the

physiognotrace supplied visual analogs for the imagined place of the Jeffersonian political subject.

The French Connection

Before delving into the representational problematics of profiles, it is useful to examine the prehistory of Hawkins's physiognotrace. Although Hawkins's device was a new invention, by 1802 the practice of making profiles already had a rich history. Further, the glowing terms in which Americans described the physiognotrace—such as "accurate," "true," and "correct"—were informed to a certain extent by rhetoric surrounding the use of similar machines in Europe.

The first-century writer Pliny the Elder attributed the art of making profiles to a Corinthian maid named Dibutades, daughter of the potter Butades. Dibutades, Pliny wrote in *Natural History,* was distraught over the impending departure of her beloved, a soldier, for battle. As the young man dozed in a chair, Dibutades memorialized his visage by outlining his shadow upon a wall. The legend became a favorite narrative of eighteenth-century painters and poets, who romanticized Dibutades's mythic tracing as the origin of drawing and design. Consider the following tribute to the British painter George Romney penned by William Haley and published, with an accompanying frontispiece engraved by Francis Kearny (figure 2.4), in a New York journal of 1808:

> Blest be the pencil! which from death can save
> The semblance of the virtuous, wise and brave . . .
> . . . Oh! Love, it was thy glory to impart
> Its infant being to this magic art!
> Inspired by thee, the soft Corinthian maid
> Her graceful lover's sleeping form portray'd
> Her boding heart his near departure knew;
> Yet longed to keep his image in her view:
> Pleas'd she beheld the steady shadow fall,
> By the clear Lamp upon the even wall:
> The line she trac'd with fond precision true,
> And, drawing, doted on the form she drew. . . .[22]

The poem concludes with Butades's transformation of his daughter's two-dimensional copy, an indexical outline of a now-absent figure, into a three-dimensional clay bust, redolent with simulated presence.[23]

Figure 2.4   Francis Kearny, "The Origin of Design," frontispiece to *The Cabinet of Genius*, New York, 1808. Courtesy of the American Antiquarian Society.

While some reimagined the first scene of "shadow painting," others revivified the antique art itself. Records date to the late seventeenth century of men and women skillfully cutting heads out of black paper. By the 1760s, the practice was attached to the name of Etienne de Silhouette, the comptroller-general in the administration of Louis XV. Silhouette, like other members of socially privileged classes, cut profiles in his leisure time. His name became synonymous with "profile" through no intent of his own; rather, Silhouette's strict fiscal policies—considered draconian in an era of exorbitant royal spending—prompted his contemporaries to describe all things spare and parsimonious as "à la Silhouette." The moniker stuck to the economical profile, austere in form and color, and eventually lost its negative connotation.[24]

Johann Caspar Lavater, a Swiss minister who began to publish his theories of physiognomy in 1772, is ordinarily credited with instigating the swell of interest in profiles that occurred during the late eighteenth century.[25] In his *Essays on Physiognomy*, written in German and translated into French and English during the author's lifetime, Lavater argued that everyone had a unique, divine essence that was externally visible in the bones of the skull. One could not detect this essence by studying faces directly, for humans could readily manipulate their soft muscular tissues to shift expression, to disguise their inner selves, and to pretend to be something other than they really were.[26] Instead, one should look for signs of internal traits in a person's shadow. "Shades," as Lavater called silhouettes, offered "the truest representation that can be given of man" and the most "immediate expression of nature, such as not the ablest painter is capable of drawing by hand."[27]

The illustrated *Essays* offered object lessons in the pseudoscience of physiognomy. Page after page, paired side by side, hundreds of profiles were presented as comparative records of character. For instance, Lavater perceived "penetration and sound understanding" in the forehead and nose of the philosopher Moses Mendelssohn, whose silhouette was reproduced in a plate with five additional profiles, but saw a "choleric, phlegmatic man" in the eyebrows and chin of an adjacent figure.[28] The *Essays* were popular in the United States, where extracts of Lavater's writings were first published in 1786, and there is evidence that people used silhouettes to ascertain knowledge of family and friends.[29] Charles Willson Peale, who conducted interpretations of profiles several times in his correspondence, clearly valued this exercise as one of the many advantages of the physiognotrace. In a letter to his daughter Angelica Peale Robinson, Peale advised against judging others by their behavior alone "because we cannot often fathom their motives; reasons may influence them which are hid from our Eyes."[30] Peale may have been thinking about Angelica's husband, Alexander Robinson, whom he disliked, for in a missive to another relative he confided that the "*head* of the *form* which nature has given to Mr. Robinson" revealed a streak of obstinacy. "I believe Lavater would pronounce the same, I will consult him the first leisure opportunity as a tryal of my judgment in Physiognomy."[31]

The source of Lavater's many physiognomic specimens was a large drawing instrument that one used to trace the projection of a shadow cast by candlelight. In an engraving of the device, as the art historian Victor Stoichita has observed, "the Plinian scenario of origins has been transformed into an actual posing session," notably one that reversed the gender roles of profiler and profiled (figure 2.5).[32] True to Lavater's belief

Figure 2.5   A Sure and Convenient Machine for Drawing Silhouettes, from *The Whole Works of Lavater on Physiognomy,* London [1787?]. Courtesy of the National Gallery of Art.

that mechanical drawing was superior to the gifts of the "ablest painter," the draftsman is largely obscured behind the instrument while the candle, in full view atop an anthropomorphic stand, performs the artist's work of making images. Stoichita suggests that Lavater's machine, with its nearly invisible delineator and a sitter pressed close to the translucent board, formed a visual counterpart of the Christian practice of confession.[33] One might also note the eerie visual and functional similarities between Lavater's device and a contemporary French contraption: the guillotine. Not unlike the latter instrument, the silhouette machine was an imposing wooden frame into which the subject physically inserted him- or herself. And in each case, the ends involved a display of the disembodied head—the Lavaterian silhouettes scrutinized for signs of interiority, the inanimate crown of the guillotine victim held up, as in one French print, before a crowd of curious spectators.[34] Charles Willson Peale made a punning equation of the two instruments in a July 1803 letter to Raphaelle, who was traveling through Virginia with his physiognotrace: "I suppose by this time you have nearly completed taking off all the heads of Norfolk."[35]

A decade before Joseph Guillotin's invention was enlisted as a tool of republican warfare, another Frenchman devised a far less traumatic means of catching heads. Gilles-Louis Chrétien's aptly named "physionotrace," depicted in a sketch of 1788 by his partner Edme Quenedey, was the immediate forerunner of the instrument that Hawkins patented in 1802 (a "g" was added to the spelling of "physionotrace" when the machine was introduced in the United States).[36] The design of the physionotrace was based on the pantograph, a collapsible mechanism of levers introduced by Christoph Scheiner in 1631.[37] At one end of the pantograph's accordionlike frame was a stylus that one used to trace the lines of a print, map, or drawing; at the other was a pencil that moved in coordination with the stylus to duplicate, reduce, or enlarge the original image. Chrétien turned Scheiner's pantograph on its side and attached it to a tall wooden structure. Peering through an eyepiece at a person seated on the far side of the physionotrace, he followed the outline of his sitter's face with a bar. As the bar moved the pantograph's arms along the vertical length of the wooden frame, a pencil attached to a lower joint of the pantograph drew the sitter's profile upon a paper mounted at the center. Like Lavater, Chrétien and his partner attributed the veracity of their silhouettes to the rapidity of their instrument. "The two minutes, at the most, that I employ for drawing the overall shape is not enough time for the model's physiognomy to change," Quenedey wrote. "From this comes the great truthfulness that one sees in all the portraits made with the Physionotrace and which astonishes the most skillful artists." Whereas Lavater's machine reproduced the shapes of shadows, however, Chrétien's device followed the positive features of a person's actual physiognomy. The sense of proximity implicit within this representational method led Quenedy to liken silhouettes to plaster masks of sitters' faces, a sculptural practice that was often heralded as an exacting method of capturing details; accomplished artists, he enthused, were comparing the silhouette portrait "to those which have been cast from life."[38]

By the 1790s, several kinds of profile machines were in use within the United States.[39] The most successful proprietor of the physionotrace was Charles Balthazar Julien Févret de Saint-Mémin, a French émigré who ran a thriving business in New York together with his countryman, Thomas Bluget de Valdenuit. Following the example of Chrétien and other physiognotrace operators, Saint-Mémin and Valdenuit engraved plates from the original life-size profile drawings, then sold sets of the drawing, plate, and prints to their clients. In 1798, Saint-Mémin moved his business to Philadelphia, where he attracted a large clientele that included political and civic luminaries, military men, and French refugees. Saint-Mémin remained active in Philadelphia until 1803, when John Isaac Hawkins's physiognotrace, the machine that guaranteed an even closer transcription of sitters, was unveiled at the Peale Museum.[40]

Citizen Silhouette

Charles Willson Peale, a radical Whig during the Revolution and a dedicated Jeffersonian in its aftermath, mounted a program of democratic hermeneutics within the frame of a political monument when he moved his museum into the Pennsylvania State House in 1802. Few early American buildings boasted such an illustrious history. Constructed in 1732 and subject thereafter to numerous additions and renovations, the State House was the colonial headquarters for Pennsylvania's elected representatives, councils, and Supreme Court. Famously, it hosted the First and Second Continental Congresses in 1774 and 1776; the Congress of the Confederation during the war; and the Constitutional Convention in 1787. In 1820, the structure was renamed Independence Hall in commemoration of the epochal event that took place in its first floor assembly room on July 4, 1776.[41]

While matters of liberty, federalism, and representation were debated in a range of early American forums—coffeehouses, newspapers, legislative bodies—the Pennsylvania State House was mythologized in paintings and prints as the locus of national identity.[42] To furnish the State House with pictures, fossils, and preserved animals, as

Peale did in 1802, was therefore to occupy a space dense with political meaning for the purposes of education, entertainment, and display. Though Peale refrained from articulating an explicitly political mission for the museum, his private vision and public promotion of the institution were thoroughly imbued with a republican ethos of egalitarianism, virtue, and civic responsibility. Peale variously styled the museum as an emporium of "useful knowledge," an "open book of nature," and a "world in miniature," thereby construing, in the words of the cultural historian Laura Rigal, "the possibility of actual or literal representation in the language of science, if not in the realm of politics or government."[43] Other scholars have pointed to the ways in which Peale's arrangement of objects in the "Long Room," the museum's principal gallery, conflated the new political order with an Enlightenment model of natural order, the "Great Chain of Being," that positioned man at the apex of a hierarchy of creation. A watercolor drawing of the Long Room made in 1823 shows portraits of republican "worthies," painted appropriately "plain," hanging above cases of stuffed birds; white neoclassical busts were likewise perched atop glass boxes of mineralogical specimens.[44] The art historian David Steinberg has observed that although this "biological approach to the display allowed viewers to find themselves represented in it," in practice common visitors to the Long Room probably found no political counterparts in the models of exemplary citizenship that stared down from the walls. "Instead," Steinberg explains, "the installation of the portraits high in the gallery offered such people a spatial and ocular metaphor of their limited social rank."[45]

Beneath the gaze of these remote virtual representatives, Hawkins's physiognotrace augured a new and improved kind of representation. Placed in the northwest corner of the Long Room, the instrument was one of many modalities of representation—visual, political, classificatory, didactic—operative within the space of the State House.[46] To have one's profile traced here, then, was to be inscribed within a pre-existing matrix of representations. Visitors to the Peale Museum took home four identical profile portraits, each a negative hollow limned by the marks of their "phyz." The positive shapes remained behind with the silhouette cutter Moses Williams. "So extensive was [Williams's] business," Rembrandt Peale recalled in the 1850s, "that I have seen two barrels full of these inner sections, which he called his *blockheads*—among which were frequently found, by careful search, the likenesses of many a valued friend or relative, and sometimes of distinguished personages—another source of profit to him."[47] These arbitrary collections of ordinary heads—the paper chaff of a representational process that rendered all sitters equal in size, shape, and color—constituted democratic assemblies in miniature, as Peale might put it. The physiognotrace allowed museum patrons to be represented visually in a manner conceptually analogous to the practice of actual representation then being articulated within contemporary political discourses. Both processes, for instance, necessarily involved a conscious decision to be represented; the act of sitting down at the machine amounted to a willing entry into representation that entailed an implicit expression of *consent,* the category of agency that political philosophers from Locke onward identified as the foundation of constitutional government and the inviolable right of republican representation.[48] There was also the matter of arranging oneself for depiction. The physiognotrace required sitters to enact a precise physical discipline—back straight, head stilled, chin up, eyes forward—that resonated with public admonishments to exercise perceptual vigilance in the political sphere. And there was the fact of tracing one's own profile, actuating self-representation, within full view of an audience of one's peers; Hawkins's instrument made visible the mechanics of representation during an era that increasingly valued political transparency.

Proprietors of physiognotraces tirelessly promoted the novelty of the physiognotrace and the infallible nature of the images it produced. "The correctness of likeness given by this engenious [*sic*] invention brings considerable numbers of Visitors to my Museum," Peale told Jefferson in June 1803.[49] In Boston, Uri K. Hill swore that his physiognotrace bested "all others in accuracy," while in Baltimore, Saint-Mémin claimed to take and engrave likenesses "in a style never introduced in this country." Indeed, one of Saint-Mémin's clients complimented "the precision of drawing and acknowledged justness of similitude" achieved by the artist's methods.[50] This rhetoric of keen resemblance, while deriving in part from enthusiasm for the sheer novelty of the physiognotrace, was also stimulated by cultural anxieties concerning the mutability of appearance. Visual likeness, as citizens of the Enlightenment were all too aware, could be a deceptive veneer. Early national essayists, novelists, and playwrights vilified charlatans of all types—demagogues, counterfeiters, ventriloquists—as figures of social and political disorder, tricksters who deceived the naive through facile manipulations of surfaces and exteriors.[51] To combat dissimulation, individuals were urged to hone their perceptual skills with the aid of optical games and illusionistic puzzles.[52] The Peale Museum, which sought to balance education with "rational recreation," featured several such "pleasing deceptions," including a set of "Magic Mirrors" that returned grossly distorted images to bemused viewers.[53] It is telling that this specular apparatus, which exposed the fallacy of a reflexive model of representation based upon assumptions of apparent likeness, was located next to the physiognotrace, a mechanical apparatus that construed likeness in terms of indexicality. The latter device supplanted the subjective, fallible judgment of visual perception with the objective proof of measured details.[54]

The representational "truth" of the profile portrait resided in its perceived status as a graphic register of individuality. Seated at the physiognotrace, the sitter mapped the physical topography of his head using the small brass gnomon (the "index") that protruded horizontally from the base of the pantograph. The device purportedly captured every facial particularity—sunken chins, swollen lips, Roman noses, rounded foreheads—and translated these details to the stylus that incised a miniature reproduction of the sitter's profile on the piece of white paper secured at the top of the board. The steely pressure of the gnomon could be unpleasant, if we can believe contemporary advertisements for other types of silhouette machines; William King, who operated a shadow instrument in Boston around 1805, informed his customers that they would "not [be] incommoded with any thing passing over [their] face." "Ladies are particularly informed," King added elsewhere, "that he takes their Profiles without their faces being scraped."[55] If Hawkins's physiognotrace was strong enough to discomfort some sitters, then it seems improbable that it would have been adequately supple to capture the slight wisps of eyelashes or unruly curls of hair visible in many silhouettes. Undoubtedly, the person who adeptly cut out the "blockheads" added these flourishes. Charles Willson Peale admitted as much in a letter to Hawkins. "The Physiognotrace is still in demand," he reported in 1807. "We contrive to give occasionally a different size, but the perfection of Moses's cutting supports its reputation of correct likeness."[56]

Adjacency was another essential aspect of the physiognotrace's representational properties. Long after the instrument had passed out of fashion, Peale remembered it as a novel means of "stamping a true likeness and character."[57] The description fittingly conveyed the physicality of accommodating one's body to a machine. It also aligned the profile with other signs of impressed or incised identity: the visual authority of a mark, the material fixity of a seal, the singular imprint of a life or death cast. Silhouettes were the products of a spatial abuttal, proof of contact between a person and a representational system. At the same time, they conveyed a temporal contiguity because they pointed back to the moment in which a sitter had pressed her head against the wooden board of the physiognotrace. Raphaelle Peale sentimentalized—even eroticized—this aspect of the profile in an advertisement that he published in Baltimore under the pseudonym "Edmund":

'Tis almost herself, Eliza's shade,
Thus by the faithful facietrace pourtray'd!
Her placid brow and pouting lips, whose swell
My fond impatient ardor would repell.

Let me then take that vacant seat, and there
Inhale her breath, scarce mingled with the air:
And thou blest instrument! which o'er her face
Did'st at her lips one moment pause, retrace
My glowing form and leave, unequall'd bliss!
Borrow'd from her, a sweet etherial Kiss.
    Such are the pleasures, Peale, attend thy art,
    To expand the finest feelings of the heart.[58]

The empty holes of the excised outlines were transformed into imagistic wholes once they were set against backgrounds of black paper and cloth. The high degree of contrast between the white paper and dark interiors created shapes that appeared to be solid. Many sitters encased their profiles in simple frames, some with decorative gilding, which Peale sold at the museum admission office for twenty-five cents.[59] Pragmatically, framing prepared the silhouette for display upon walls at home; visually, it secured an iconic presence for the indexical tracing, materializing the fugitive profile likeness. Other sitters chose to exchange silhouettes as gifts, much like watercolor miniatures, or to paste them into books as pictorial documents of marriages and genealogies, as in a family album that features numerous "blockheads" alongside profiles mounted on blue-painted paper (figure 2.6).[60] Often, sitters further customized their profiles by recovering signs of personal identity that the physiognotrace or Williams's scissors elided: Shirts and collars, combs and ribbons, and coils of hair appear sketched around the edges of many silhouettes.

Figure 2.6   "Blockhead" and profiles pasted in the Collins family album of silhouettes. Courtesy of the Library Company of Philadelphia.

For all the perceived merits of the machine-made profile, this was decidedly *not* the proper way to obtain a portrait during the early nineteenth century. Neoclassical art theory emphatically disapproved of "particular" methods of representation. Taught that their aim was to portray a sitter's elusive internal nature or "character," artists were urged to overlook unseemly physiognomic details. "The likeness of a portrait," Sir Joshua Reynolds explained in one of the *Discourses* he delivered as president of the Royal Academy, "consists more in preserving the general effect of the countenance, than in the most minute finishing of the features, or any of the particular parts."[61] Portraitists were instructed to use their creative faculties of imagination and invention to evoke this "general effect." Frequently, however, they communicated aspects of their sitters' character by depicting them with conventionalized emblematic attributes or by casting them in stock poses, costumes, and gestures.

By contrast, profile portraits, which privileged the contours of the head as the only authentic signs of interiority, literally effaced the need for surrounding visual cues and, just as important, undermined the role of the artist. By enabling an individual to trace his own profile, the physiognotrace reconfigured the triangular operation of painting portraits—in which sitter and portraitist together produced a third entity, the image of the sitter—as a linear equation in which the sitter enjoyed a supremely direct relation to his self-made representation. There was already some hint of this transformation in the print of Lavater's silhouette machine, which obscured the artist behind the instrument and sitter. But Hawkins's physiognotrace effectively merged the subject and act of representation, collapsing signified and signifier. Profile portraits were ontologically inseparable from their referents, spatially and temporally bound to real bodies. This sense of doubling extended to the cellulose materiality of the profile; paper, flat and sheer, was metaphorically akin to skin.

The Limits of Representation

Direct, particular, proximate—these were the keywords of actual political representation. As the historian Gordon Wood has explained, the "natural aristocracy" envisioned by the Federalists—the Constitutional framers who rose to power during the 1780s—was at odds with the American public's long-standing faith in the power of local political participation. Since the seventeenth century, many American communities had experimented with direct self-legislation in town meetings and colonial assemblies. At the close

of the eighteenth century, Federalists' efforts to structure a strong, centralized bureaucracy aggravated concerns that citizens were losing control over their familiar systems of
political representation. State legislatures were thought to have become dangerously detached from the people's control, and voters increasingly scrutinized their candidates and
delegates for signs of dissimulation. This growing mistrust of government fueled debates
surrounding the ratification of the Constitution and sparked an Antifederalist "defense of
the most localist and particularist kind of representation," writes Wood, "an actuality of
representation that was much more in accord with America's social and political realities
than the sort of representation advocated by the Federalists."[62] During the 1790s, events
including the French Revolution and Federalist abuses of power further fueled conviction
in the merits of democratic or actual representation. The voters who elected Jefferson to
power in the "Revolution of 1800" wanted delegates who closely resembled their constituents, and they measured that likeness in terms of shared interests, geographical proximity, and personal familiarity. Citizens sought tighter relationships with their elected
representatives by subjecting them to new residency requirements and issuing binding instructions in writing. Nearness of place implied similarity of mind (and facilitated surveillance of a representative's activities, thwarting potential misrepresentation).

In such a climate of "electoral narcissism," as the literary critic Mark P. Patterson aptly
puts it, one elects individuals in whom one sees oneself, those who look like or—and
here was the ever-present danger—deceptively alter their appearances to make themselves look like the voter.[63] Patterson's comment suggests the hopeful, earnest nature of
actual representation. Indeed, as historians have emphasized, actual representation was
an experimental project of vast geographical proportions, not a pat, seamless theory. In
theory, actual representation proposed a radical reorganization of the political system; in
practice, it proved difficult to implement and sustain, hampered by the realities of a
growing populace, expanding territories, and competing political interests.

Much as direct representation was an idealistic political model, the physiognotrace
was an imperfect representational index; it was contingent upon the sitter assuming a
perfect posture, unable to capture certain characteristic marks (such as scars), and only
as good as its mechanical upkeep. The machine reduced personhood to the sum of its
skeletal proportions, and in doing so it revealed more about Enlightenment rationalism
than it did about the subjectivity of individual sitters. In more subtle ways, this much
lauded instrument of correct depiction also exposed the practical limits of actual representation. The profiles inscribed on paper at the top of the pantograph could never

become the exact copies that they were claimed to be for the simple reason of *scale*. Not unlike the Peale Museum, a cloistered cabinet of wonders that purported to represent the entire world in miniature, the miniature silhouettes that one took home from the museum were merely token filaments of a partial, lineated self. Similarly, the fantasy of self-actualized and direct access to representation could be sustained only by ignoring the creative agency of Moses Williams and other cutters of profiles. Williams was the hidden artist in the Peale Museum's Long Room; although sitters might be able to trace their own profiles, they relied upon Williams's skill and intervention to complete the process and deliver the finished product. Finally, the urgent tenor in which the physiognotrace was persistently touted suggests the strain of a rhetoric born of overdetermination. Accurate, true, like, correct, direct: Reiterated ad nauseam, these terms masked the uncertainty of attaining actual representation.

There are ample indications that Peale and his contemporaries sensed limitations in the physiognotrace's capacity for exact representation. "I have to observe, that many trials will be necessary before you take a complete profile," William Pinchbeck advised potential users of his physiognotrace. "Much depends on place [*sic*] the person in a proper attitude, or perhaps you will suffer the tracing point to err from the line of the face."[64] Peale experienced similar frustration in his efforts to obtain an accurate likeness of Jefferson. Shortly after acquiring Hawkins's physiognotrace, Peale began to make tracings of Jefferson's profile using a plaster replica of a marble bust that had been modeled by the French sculptor Jean-Antoine Houdon in 1789. Though he was not entirely pleased with the quality of the resulting silhouettes, Peale generated several hundred profiles and selectively distributed them to museum visitors.[65] This procedure of making profiles from a plaster bust—a copy that was already twice removed from the original sitter—would appear to contradict the desired ideal of actuality. However, the practice was sanctioned by none other than Lavater himself, who made silhouettes of the Apollo Belvedere from a cast of the venerated classical sculpture.[66] Still dissatisfied with the profiles in June 1804, Peale wrote to Jefferson on behalf of Raphaelle and Rembrandt to request an up-to-date silhouette of the president's appearance:

> My Sons here are very desireous of having the outline of your profile taken in the size which the argand Lamp will give on the wall, resting the head against a beer or some other long Glass. It will only be the loss of one minute of time to sit, while Mr Burrill might trace the shadow with a Pencil, from which my Son Raphaelle can cut it out, and with his Physiognotrace produce numbers. The Profiles I have generally seen of you are not so critically like as I would wish them. But the best apology I can make for asking this

favour, is, that my Sons desire to present them to every visitor of the Exhibition of the Skeleton, which will be a gratification to your numerous friends.[67]

Jefferson acquiesced and forwarded two profiles to Philadelphia. Peale was sorely disappointed with the outcome. "They are so miserably done that I have had my doubts whether I ought to send them," he complained to Raphaelle, "however with a profile taken from Houdon's bust, which I send, you may perhaps make something of a Likeness, but whether of sufficient accuracy [to do] you credit I cannot say."[68]

It is not entirely clear whether Peale was advising Raphaelle to reproduce Jefferson's profile using the best of the prototypes or whether he was suggesting that Raphaelle should combine the silhouettes in order to "make something of a Likeness," however compromised in "accuracy." If Peale intended that Raphaelle should follow the latter recommendation, then he anticipated by many decades the emergence of the composite photograph. During the 1880s, the Englishman Francis Galton, best known as the father of eugenics, superimposed photographic negatives depicting the faces of felons and thieves in an attempt to develop a visual typology of criminality.[69] Galton's experiments took shape in a period that replaced physiognomy with another pseudoscience of the head, phrenology, and that enlisted the developing techniques of photography in the service of bureaucratic classification and social policing. The age that gave rise to the mug shot enabled "profile" to attain an ominous legalistic meaning: "profiling."

Isolated problems with the physiognotrace, however, did little to dissuade popular faith in the truth value of mechanical depiction. Any perceived shortcomings of machine-made portraits were certainly long forgotten by the advent in the late 1830s of a new process of imaging that once again promised unprecedented degrees of representational veracity and precision: the daguerreotype. Just as line had reproduced the shape of a sitter's skull, now light, the "pencil of nature," fixed the sitter's appearance on small silvered plates. And where the profile had been celebrated as actual representation realized, so the daguerreotype was associated with popular political ideals. In his first letter to Jefferson about the physiognotrace, Peale had coupled Hawkins's political persuasion with the man's aptitude for inventing representational technologies: "His republicanism is our only securiety [sic] for possessing of him." If Peale, a patriot of the founding generation, has the opening word in this essay, then it seems appropriate to give the last one to a democrat of the succeeding era: The daguerreotype, wrote Ralph Waldo Emerson in 1841, makes "the artist stand aside and lets you paint yourself," and therefore was "the true Republican style of painting."[70]

Acknowledgments

This essay is extracted from a chapter of the author's Ph.D. dissertation, "Likeness and Deception in Early American Art." Previous versions of the paper were presented at *Media in Transition*, a conference organized by the Massachusetts Institute of Technology in October 1999, and at a colloquium held at the Omohundro Institute of Early American History and Culture in April 2001. I thank the participants at both events for their questions and suggestions. I am especially grateful to Lisa Gitelman for her encouragement and intellectual generosity. Research for this study was generously funded by an Andrew W. Mellon Foundation Fellowship at the Library Company of Philadelphia and a Wyeth Fellowship at the Center for Advanced Study in the Visual Arts, National Gallery of Art.

Notes

1. Peale's efforts to secure the use of the State House from the Pennsylvania legislature are documented in *The Selected Papers of Charles Willson Peale and his Family,* 5 vols., ed. Lillian B. Miller (New Haven: Yale University Press, 1983–), 2:380–418 (hereafter abbreviated *Selected Papers*). See esp. "Excerpt: *Minutes of the Senate of the Commonwealth of Pennsylvania* (Lancaster, 16 February 1802)," 2:400–401.

2. *Aurora,* 28 December 1802, reprinted in ibid., 2:478. Emphasis added. The editors of the *Selected Papers* note that the advertisement also appeared on the same date in the *Gazette of the United States*. A reconstruction of Hawkins's physiognotrace is on view in the Second Bank of the United States, Philadelphia. Secondary studies of the instrument include (in order of publication): Edgar P. Richardson, Brooke Hindle, and Lillian B. Miller, *Charles Willson Peale and His World* (New York: Harry N. Abrams, 1983); Ellen Miles, *Saint-Mémin and the Neoclassical Profile Portrait in America* (Washington, D.C.: A Barra Foundation book copublished with the National Portrait Gallery, Smithsonian Institution, 1994), 106–113; David Brigham, *Public Culture in the Early Republic: Peale's Museum and Its Audience* (Washington, D.C.: Smithsonian Institution Press, 1995), 68–82; Peter Benes, "Machine-Assisted Portrait and Profile Imaging in New England after 1803," in "Painting and Portrait Making in the American Northeast," *Dublin Seminar for New England Folklife Annual Proceedings* 19 (Boston, 1996): 138–150; Miles, "1803—The Year of the Physiognotrace," in ibid., 118–137; and Edward Schwarzchild, "From the Physiognotrace to the Kinematoscope: Visual Technology and the Preservation of the Peale Family," *Yale Journal of Criticism* 12, no. 1 (1999): 57–71.

3. Charles Willson Peale to Thomas Jefferson, Philadelphia, 10 January 1803, in *Selected Papers* 2:480–482. Peale reported to Jefferson, "My progress to improvements and dress of the Museum goes on faster than I had expected—the utility of it will be rendered more conspicuous with this winters labour."

4. Jefferson to Peale, Washington, D.C., 23 January 1803, in ibid., 2:482–483; and Peale to Jefferson, Philadelphia, 28 January 1803, in ibid., 2:483–484. Jefferson wrote: "I was struck with the notice in the papers of mr Hawkins's physiognotrace, of the work of which you send me some specimens, which I perceive must have been taken from Houdon's bust. When you shall have nothing else to do I would thank you for an explanation of the principle of it, for I presume no secret is made of it as it is placed in the Museum." The editors of the *Selected Papers* suggest that Peale probably used David Rittenhouse's plaster copy of a marble bust of Jefferson, which Jean-Antoine Houdon had sculpted in 1789. In his letter of January 28th, Peale informed Jefferson that Hawkins had also "contrived another Index" designed to portray a sitter in three-quarters profile. He included a rudimentary line drawing that suggested the device required a sitter to insert his head into a square frame. "This kind of Index" was "not fitted for a public Museum," Peale explained, because it necessitated more delicate handling than Hawkins's other physiognotrace.

5. Brigham, *Peale's Museum,* 70–71.

6. Peale to Rembrandt and Rubens Peale, Philadelphia, 1 April 1803, in *Selected Papers* 2:517–518.

7. Malbone listed "a pentagraph and a small box" among the objects he temporarily deposited in the care of a friend around 1805. See "Account book and register of portraits, 1794–1807, of Edward Green Malbone," Downs Collection of Manuscripts and Ephemera, Winterthur Museum.

8. On Greenwood, see Georgia Brady Barnhill, "'Extracts from the Journals of Ethan A. Greenwood': Portrait Painter and Museum Proprietor," *Proceedings of the American Antiquarian Society* 103, no. 1 (1993), 91–178. See Miles, "1803—The Year of the Physiognotrace," 135–136, regarding the individuals to whom Hawkins extended his patent; and Benes, "Machine Assisted Profile and Portrait Imaging," on the proliferation of physiognotraces in New England.

9. Raphaelle Peale advertised the display of "ten thousand profiles taken from a collection of three hundred and forty-five thousand seven hundred and twenty" in the *Boston Gazette,* September 13 and 20, 1804. See *Selected Papers* 2:750–751 for a reprint and additional documents regarding Raphaelle's use of the physiognotrace.

10. Peale to John Isaac Hawkins, Philadelphia, 17, 22, and 25 December 1805, in *Selected Papers* 2:916.

11. Benes similarly observes that "a curious transformation was taking place in which the mechanical accuracy of portrait making seemingly eclipsed the value of the portrait itself." He attributes this change to "the pressure of entrepreneurial promotion." Benes, "Machine-Assisted Imaging," 146.

12. See, for instance, Peale to Jefferson, Philadelphia, 10 January 1803, in *Selected Papers* 2:482; Peale to Jefferson, Philadelphia, 2 June 1803, in ibid., 2:532–534; and Peale to Hawkins, 17 May 1807, Philadelphia, in ibid., 2:1014.

13. "Mammoth," *Federal Gazette and Baltimore Daily Advertiser,* Baltimore, 26 May 1804, in ibid., 2:677.

14. "Facietrace," *Boston Gazette,* 13 September 1804, in ibid., 2:751.

15. In developing these ideas, this essay charts a different tack than many prior studies of early American imagery. Art historians have successfully articulated the ways in which paintings, sculpture, and other art objects advanced the ideologies of revolution and republicanism. Less often have they considered the potential political import of images that seem apolitical in subject, style, or iconography. With this approach, I take a cue from Jay Fliegelman's description of an "interactive" Revolutionary culture in which "political ideas are present in nonpolitical texts and vice versa." Fliegelman, *Declaring Independence: Jefferson, Natural Language, and the Culture of Performance* (Stanford: Stanford University Press, 1993), 4.

16. Christopher Prendergast, *The Triangle of Representation* (New York: Columbia University Press, 2000), 4–6. The classic study of this topic is Hanna Pitkin, *The Concept of Representation* (Berkeley: University of California Press, 1967).

17. W. J. T. Mitchell, "Representation," in *Critical Terms for Literary Study,* ed. Frank Lentricchia and Thomas McLaughlin (Chicago: University of Chicago Press, 1990), 11–22.

18. Richard Bernheimer incisively details the operations of legal representation (a category into which he places political representation) in *The Nature of Representation: A Phenomenological Inquiry* (New York: New York University Press, 1961), esp. 108–124.

19. On models of virtual and actual political representation in Revolutionary and post-Revolutionary America, see Gordon Wood, "Representation in the American Revolution" (Charlottesville: Published for the Jamestown Foundation of the Commonwealth of Virginia by the University Press of Virginia, 1969); Jack R. Pole, *Political Representation in England and the Origins of the American Republic* (Berkeley: University of California Press, 1971); and John Phillip Reid, *The Concept of Representation in the Age of the American Revolution* (Chicago: The University of Chicago Press, 1989).

20. Prendergast, *The Triangle of Representation,* 6.

21. Mitchell, "Representation," 11–12.

22. Haley's poem was reprinted in *The Cabinet of Genius; containing all the theory and practice of the fine arts* (New York: Thomas Powers, 1808), 21–22. The concluding section reads: "Nor,——as she glow'd with no forbidden fire/Conceal'd the simple Picture from her sire/His kindred fancy, still to nature just/Copied her line, and form'd the mimic bust./Thus from thy power love, we trace/The Modell'd image, and the Pencill'd face!"

23. On the popularity of the narrative of the Corinthian maid during the eighteenth-century, its significance for the ideological foundations of art history, and its implicit problematics of gender and creativity, see Ann Bermingham, "The Origin of Painting and the Ends of Art: Wright of Derby's *Corinthian Maid,*" in *Painting and the Politics of Culture: New Essays on British Art 1700–1850,*

ed. John Barrell (Oxford: Oxford University Press, 1992), 135–164; and Richard Shiff, "On Criticism Handling History," *History of the Human Sciences* 2, no. 1 (1989): 63–87. Other notable discussions of the subject include Jacques Derrida, *Memoirs of the Blind* (Chicago: University of Chicago Press, 1993), 49–51; George Levitine, "Addenda to Robert Rosenblum's 'The Origin of Painting: A Problem in the Iconography of Romantic Classicism,'" *Art Bulletin* 40 (December 1958), 329–331; Robert Rosenblum, "The Origin of Painting: A Problem in the Iconography of Romantic Classicism," *Art Bulletin* 39 (December 1957), 279–290; Mary D. Sheriff, "Art History: New Voices/New Visions," *Eighteenth-Century Studies* 25, no. 4 (summer 1992): 427–434; and Victor Stoichita, *A Short History of the Shadow* (London: Reaktion Books, 1997), 123–127, and 153–155.

24. Garry Apgar, "Silhouette," in *The Dictionary of Art,* ed. Jane Turner, vol. 28 (New York: Grove's Dictionaries, 1996), 713–714. For a survey of the production and practices of profile portraiture between the late seventeenth and nineteenth centuries, see E. Nevill Jackson, *The History of Silhouettes* (London: The Connoisseur, 1911).

25. Johann Caspar Lavater, *Von der Physiognomik* (Leipzig, 1772). The first illustrated work followed in 1775: *Fragmente—Physiognomische Fragmente, zur Beförderung der Menschenkenntniss und Menschenliebe,* 4 vols. (Leipzig, 1775–1778). French and English translations appeared, respectively, in 1781 and 1788–1789. See Joan K. Stemmler, "The Physiognomical Portraits of Johann Caspar Lavater," *Art Bulletin* 75, no. 1 (March 1993): 151–168, for full citations of the many editions of Lavater.

26. On these points see Stemmler, "Johann Caspar Lavater," 156–157; and Stoichita, *A Short History of the Shadow,* 159. See also Barbara Maria Stafford, *Body Criticism: Imaging the Unseen in Enlightenment Art and Criticism* (Cambridge, Mass.: The MIT Press, 1991), 84–103; and idem, "'Peculiar Marks': Lavater and the Countenance of Blemished Thought," *Art Journal* (fall 1987), 185–192.

27. Lavater, *Essays on Physiognomy; for the Promotion of the Knowledge and the Love of Mankind . . .* (Boston: Printed for William Spotswood & David West, 1794), 218.

28. Ibid., 221–222.

29. Excerpts of the English translation first appeared in the United States in "Extracts from a Treatise on Physiognomy," *New Haven Gazette, and the Connecticut Magazine* 1, no.41 (23 November 1786): 318–319. Cited in Brandon Brame Fortune, "Portraits of Virtue and Genius: Pantheons of Worthies and Public Portraiture in the Early American Republic, 1780–1820" (Ph.D. dissertation, University of North Carolina at Chapel Hill, 1986), 218. The full edition of Lavater was published in the United States in 1794 (see note 27).

30. Peale to Angelica Peale Robinson, Philadelphia, 17 March 1803, in *Selected Papers* 2:515–516.

31. Peale to Nathaniel Ramsay, Philadelphia, 17 March 1805; in ibid., 2:816. Emphasis in original. See also Peale to Deborah Jackson, Philadelphia, 8 January 1807, in ibid., 2:998–999.

32. Stoichita, *A Short History of the Shadow,* 156.

33. Ibid., 164.

34. See the engraving of Louis XVI's beheading reproduced in Lynn Hunt, *The Family Romance of the French Revolution* (Berkeley: University of California Press, 1991), 9. Hunt indicates that the engraving originally appeared in *Révolutions de Paris,* no. 185 (19–26 January 1793).

35. Peale to Raphaelle Peale, Philadelphia, 19 July 1803, in *Selected Papers* 2:583. Peale reprised the joke in a letter of February 9, 1805 to Rembrandt Peale: "I suppose you have seen a parragraph [*sic*] of a Baltimore paper, which says that 'Rembrandt Peale is taking off the heads of the Members of Congress for his Museum at Philada.' It would have been much better to have, to give the Printers the true state of the Pun, <*which*> as it is of the first grade of Puns." See ibid., 2:800.

36. Quenedey's sketch of Chrétien's *physionotrace* is reproduced in Miles, *Saint-Mémin,* 42.

37. Christoph Scheiner, *Pantographice* (Rome: Ex Typographia Ludouici Grignani, 1631). The pantograph was widely employed as a copying and "perspective machine" during the seventeenth and eighteenth centuries. For further information about its history, form, and uses, see Maya Hambly, *Drawing Instruments 1580–1980* (London: P. Wilson for Sotheby's Publications, 1988); and Martin Kemp, *The Science of Art: Optical Themes in Western Art from Brunelleschi to Seurat* (New Haven: Yale University Press, 1990). Robert Dossie's *The Handmaid to the Arts,* a popular artist's manual (and a copy of which Peale owned), pronounced the device an aid "to those who have no facility in drawing" and supplied illustrated directions for those who wished to construct it. See Dossie, *The Handmaid to the Arts* (London, 1758), 345–349.

38. Quoted in Miles, *Saint-Mémin,* 44. See pp. 43–45 for further information about Chrétien, Quenedey, and examples of their work.

39. The first secure documentation of a mechanical instrument resembling Chrétien's *physionotrace* (as opposed to a Lavaterian shadow device) dates to 1796, when a French artist named J. J. Boudier installed his machine at 275 Front Street, Philadelphia. See Miles, *Saint-Mémin,* 64–65, regarding Boudier's practice in Maryland and Pennsylvania.

40. See Miles, *Saint-Mémin,* 85–113, on Saint-Mémin's tenure in Philadelphia and the competition he faced within the city and surrounding region from physiognotrace operators in addition to Hawkins.

41. On the history of the State House, see Frank M. Etting, *An Historical Account of the Old State House of Pennsylvania, Now Known as the Hall of Independence* (Boston: James R. Osgood & Co., 1876); J. Thomas Scharf and Thompson Westcott, *History of Philadelphia, 1609–1884,* vol. 3 (Philadelphia: L.H. Everts & Co., 1884); Penelope Hartshorne Batcheler, *Independence Hall Historic Structures Report* (Philadelphia: Independence National Historical Park, 1992); and Charlene Miers, "Slavery, Nativism, and the Forgotten History of Independence Hall," *Pennsylvania History* 67, no. 4 (autumn 2000): 481–501.

42. Several eighteenth-century illustrations of the structure are reproduced in Edward M. Riley, "The Independence Hall Group," in *Historic Philadelphia: From the Founding until the early Nineteenth*

Century, Transactions of the American Philosophical Society 43, no. 1 (1953): 7–42. William Birch included two prints depicting the State House in his *Eleven of the Principal Views of Birch's Philadelphia* (Philadelphia, 1800). Depictions of the building's interior in paintings include Robert Edge Pine and Edward Savage, *Congress Voting Independence* (1785), and John Trumbull, *The Declaration of Independence 4 July 1776* (1787–1820).

43. Laura Rigal, *The American Manufactory: Art, Labor, and the World of Things in the Early Republic* (Princeton, N.J.: Princeton University Press, 1998), 96.

44. For a color reproduction of the Long Room watercolor drawing, see Richardson, Hindle, and Miller, *Charles Willson Peale and His World,* 82–83. On Peale's pantheon of republican "worthies" and the political significance of his representational "plain style," see Fortune, "Portraits of Virtue and Genius"; and idem, "Charles Willson Peale's Portrait Gallery: Persuasion and the Plain Style," *Word & Image* 6, no. 4 (October–December 1990): 308–324.

45. David Steinberg, "Charles Willson Peale Portrays the Body Politic," in *The Peale Family: Creation of a Legacy, 1770–1870,* ed. Lillian B. Miller (Published by Abbeville Press in association with The Trust for Museum Exhibitions and the National Portrait Gallery, Smithsonian Institution, 1996), 131.

46. According to Peale's *Guide to the Philadelphia Museum* (Philadelphia, 1804), the instrument was located in the Long Room: "A person attends in this room with Hawkins' ingenious Physiognotrace, for the purpose of drawing Profiles." A footnote added, "The Attendant is allowed to receive 8 Cents for cutting out each set of Profiles, from such as choose to employ him." See *Selected Papers* 2:763. See also Charles Coleman Sellers, *Charles Willson Peale* (New York: Charles Scribner's Sons, 1969), 334 and 341, which reproduces a floor plan of the museum that locates the physiognotrace in the northwest corner of the Long Room.

47. Rembrandt Peale, "Notes and Queries. The Physiognotrace," *Crayon* 4, part 10 (October 1857): 308.

48. On the issue of consent, see esp. Reid, *The Concept of Representation in the Age of the American Revolution.*

49. Peale to Jefferson, Philadelphia, 2 June 1803, in *Selected Papers* 2:534. Peale's descriptions of the polygraph, a writing instrument that was related in form and function to the physiognotrace, further reveal his ideas about accurate representation. In advertisements and correspondence with Hawkins and Jefferson, with whom he developed the polygraph, Peale touted the polygraph's capacity to produce "true," "fac simile," and "correctly alike" copies of letters in multiple. See for instance Peale to Hawkins, Philadelphia, 3 March 1807, in ibid., 2:1005. For further information about the polygraph, see Silvio Bedini, *Thomas Jefferson and his Copying Machines* (Charlottesville: The University Press of Virginia, 1984).

50. Hill is cited in Benes, "Machine-Assisted Imaging," 138; for Saint-Mémin, see Miles, *Saint-Mémin,* 116 and 124.

51. On early national fears of deception, see Richard Hofstadter, *The Paranoid Style in American Politics and Other Essays* (New York: Knopf, 1965). Wood emphasizes that the problem of deception was a broadly eighteenth-century Anglo-American preoccupation in "Conspiracy and the Paranoid Style: Causality and Deceit in the Eighteenth Century," *William and Mary Quarterly,* 3d ser., 39 (1982): 401–441. For a study that relates early American literature to cultural concerns about deception, see Mark P. Patterson, *Authority, Autonomy, and Representation in American Literature, 1776–1865* (Princeton: Princeton University Press, 1988).

52. See Barbara Maria Stafford, *Artful Science: Enlightenment Entertainment and the Eclipse of Visual Education* (Cambridge: MIT Press, 1994).

53. Brigham, *Peale's Museum,* 81–82, quotes from the diary of Catherine Fritsch, who compared the physiognotrace and Magic Mirrors. "Whereas the mirrors distorted one's self-perception," Brigham comments, "the silhouette machine fixed an accurate portrayal on paper."

54. Notably, Chrétien lauded the likenesses achieved by his physionotrace over those produced by instruments fitted with optical lenses. Chrétien boasted that even at a distance of fifteen feet, a draftsman "would be close enough to the person he is drawing to be able to distinguish and copy every hair . . . without the aid of an optical glass." Quoted in Miles, "1803—The Year of the Physiognotrace," 128.

55. King is quoted in Benes, "Machine-Assisted Imaging," 141.

56. Peale to Hawkins, Philadelphia, 17 May 1807, in *Selected Papers* 2:1014.

57. Peale, *Autobiography,* in *Selected Papers,* 5:310. In "1803—The Year of the Physiognotrace," Miles emphasizes that the physiognotrace enjoyed terrific success upon its introduction in 1802–1803 and remained popular for several years. Peale observed a significant lull in business as early as May 1804—"it at present appears to be pritty well *done over,*" he informed Hawkins—although his collection of profiles continued to grow, numbering eight thousand by 1806. See Peale to Hawkins, Philadelphia, 22 April and 2 May 1804, in *Selected Papers* 2:659. The quantity of profiles made at the museum is a matter of some confusion. In his *Autobiography,* written in 1825–1826, Peale wrote that over 8,880 profiles were made in 1803, the physiognotrace's first year at the museum (see ibid., 5:310). Scholars have not questioned the discrepancy between this claim and a statement that Peale made in 1806 that suggested that 8000 profiles were made in the three years *between* 1803 and 1806. See Peale to Jefferson, Philadelphia, 13 December 1806, in ibid., 2:991.

58. "Edmund" [Raphaelle Peale], "On seeing Eliza's Profile, drawn with the facietrace, by Mr. Peale," *Federal Gazette and Baltimore Daily Advertiser* (13 June 1804); reprinted in *Selected Papers* 2:710.

59. Brigham, *Peale's Museum,* 71.

60. On the social uses of profiles, see ibid., 68–82; and Anne Verplank, "Facing Philadelphia: The Social Functions of Silhouettes, Miniatures, and Daguerreotypes, 1760–1860" (Ph.D. dissertation, College of William and Mary, 1996).

61. Joshua Reynolds, *Discourses on Art,* ed. Robert R. Wark (New Haven: Published by the Paul Mellon Centre for Studies in British Art (London) Ltd. by Yale University Press, 1975), 259.

62. Wood, "Representation in the American Revolution," 46.

63. Patterson, *Authority, Autonomy, and Representation in American Literature,* 44–45.

64. Pinchbeck describes his instrument in *Witchcraft: or the Art of Fortune-Telling* (Boston, 1805); reprinted in Benes, "Machine-Assisted Imaging," 150.

65. "Several hundreds taken from Houdons Bust I have dispersed, altho' I keept [*sic*] them out of sight, and meant to be a favor. The size of that profile was not so good for framing, as my Sons Physiognotrace gives," Peale told Jefferson in letter of June 15, 1804. See *Selected Papers* 2:711.

66. See Stoichita, *A Short History of the Shadow,* 165–167.

67. Peale to Jefferson, Baltimore, 15 June 1804, in *Selected Papers* 2:711. "Mr. Burrill" was Jefferson's secretary, William Armisted Burwell.

68. Peale to Raphaelle Peale, Philadelphia, 23 June 1804, in ibid., 2:720.

69. See Allan Sekula, "The Body and the Archive," *October* 39 (winter 1986): 3–64.

70. Quoted in Alan Trachtenberg, *Reading American Photographs: Images as History, Mathew Brady to Walker Evans* ([New York]: Hill and Wang, 1989), 29.

3 Children of Media, Children as Media: Optical Telegraphs, Indian Pupils, and Joseph Lancaster's System for Cultural Replication

Patricia Crain

Everywhere I turned I found a "squared world," a society so compartmentalized that life, including my own, had no room to move around. . . . I unwittingly internalized it—tore my life-web and stuffed the broken strands into the "boxes. . . ."
—Marilou Awiakta, Cherokee, 1997[1]

We have got to become men and women and we have got to take our place in line in life. . . . You have got to march through this world; the world expects you to do something, not simply to play and not simply to have pleasure.
—Richard Ballinger, secretary of the interior, addressing Indian students, 1909[2]

A PLACE FOR EVERY THING AND EVERY THING IN ITS PLACE." Joseph Lancaster's motto penetrated hundreds of schoolrooms in the United States and across the British Empire, Western Europe, Russia, and Latin America between 1800 and 1850. Although the slogan has the texture of proverb, Lancaster coined the phrase, cannily translating Enlightenment rhetoric into the language of the schoolhouse.[3] In its echo of the "places"—*loci* or *topoi*—of rhetorical invention, the expression conveys a technique for organizing knowledge and transmitting data. In this sense the slogan begins to map the place of Lancasterian pedagogy among the media technologies of the early nineteenth century, including not only the codex but also the optical telegraphs built, planned, or imagined by his contemporaries. In its tautology, Lancaster's motto captures the self-sufficiency of his system and its effacement of the power dynamic that motivates the placing (and displacing) of people and things. In this sense the phrase expresses the consonance between Lancaster's pedagogy and the discourse of colonialism.[4]

Originally fashioned to teach the London poor, Lancaster's "monitorial" system was widely adopted for missionary projects, particularly those of the American Board of Commissioners for Foreign Missions. And in 1821 the Bureau of Indian Affairs, then a branch of the war department, promoted it specifically for teaching American Indians.[5] That a pedagogy designed for the urban poor became a boon to missionaries on the southern frontier reveals the striking similarities between the child produced by Lancaster's system and the American Indian imagined by colonial discourse. It is not simply that the Indian was the "child of the forest," or that poor children and Indians both required the supplement of "civilization." Rather, the representational techniques and technologies in Lancaster's manuals and classrooms, which "place" students and which create a template for their uncanny replication, translate in full, if with delicate calibrations, to the mission project.

Joseph Lancaster was motivated by a missionary impulse from the start. Born in the artisan class in London in 1778, Lancaster read an abolitionist essay by Thomas Clarkson that determined him at age fourteen to go to Jamaica to teach African slaves.[6] The story goes that he put out to sea with a Bible and *Pilgrim's Progress* in his kit but was returned to his family after three weeks. In 1798, young and poor himself, Lancaster set up a school for the poor, supported by Quakers, his adopted sect. In part a figure of the late Enlightenment, in its missionary and Benthamite molds,[7] Lancaster is in part a figure of romanticism, a blowzy Byronic character. Profligate, spendthrift, overweight, paranoid, probably a sadist, possibly a pedophile, half altruist, half self-promoting snake-oil salesman, Lancaster was a ready-made American. Indeed, he was hounded out of England and was embraced by the United States, which, by the time Lancaster arrived in 1818, had instituted an estimated 150 schools according to his plan, from New York and Philadelphia, to Tennessee and North Carolina, to Cincinnati and Detroit.[8]

In the early national United States, as in England, pedagogical theory was haunted by the French Revolution, the monitory example of the revenge of the unalphabetized.[9] While the United States didn't have a so-called "peasant class," it did have, along with its growing slave, free black, and indigenous populations, a burgeoning white population whose class affiliations were worryingly indeterminate. For while the United States warmly promoted its "rising generation," it was a tide that rolled in a little too fast for comfort: The 1790 census found that 49 percent of the white population was under 16, which was also the median age of Americans from the 1790s to 1830.[10] How to cope with such an immigration from within? How to incorporate and acculturate all of these "little strangers"?

Lancaster's system promised wholesale acculturation and social control at discount prices.[11] As Carl Kaestle notes, however, it was not only the outcome that appealed to reformers but the very idea of method applied to the training of youth, an ambition of theorists of the republic from Thomas Jefferson to Noah Webster to Benjamin Rush, among many others. Lancaster offered a

> fundamentally new idea in education, informed by the image of the factory, the reality of technological advance, the incipient growth of bureaucracy in response to demographic growth, and, especially, a desire for order in response to the increasingly obvious threat of chaos in the lives of the working classes. The answer was a system that was the essence of technology: it was not simply efficient, as a single machine is, but, like the machine, it was infinitely replicable.[12]

As Kaestle suggests, much of the promise of the system was embedded in the very terms of its promotion, communicating the soothing notion that rational principles of mechanization and manufactory could apply to the thorny problem of education, for which organic metaphors of cultivation had long been the mainstay. The new terminology was taken to like a tonic and its result seemed miraculous. Lancaster could cheaply instruct hundreds of children at a time; the school he established in London could tutor 500 students, from the ABCs on up, in one great room.

Like a good manufactory, the Lancaster school could turn out product.[13] And as if to assure his clientele of his expertise at replication, Lancaster promoted his system through the repetition of images supplementing his texts. Lancaster's illustrated manuals are blueprints for pedagogues and school committees. The illustrations, along with specs for materials, buildings, furnishings, and apparatus, give his manuals a can-do aura; they are, in the new technological sense, "plans." What they map is the placement of children. The Lancasterian manual represents children arrayed in classrooms—standing in groups, performing tasks, circulating through space. In these images, there are almost always two notable truants. The first is the absent teacher (figure 3.1). Lancaster's revolutionary management trick is, after all, simply to delegate; the innovation that he shared with his British rival, Andrew Bell, is the "mutual" or "monitorial" (Lancaster's term) positioning of more experienced students as instructors, boys teaching other boys, in small groups organized by level of accomplishment.

The second truant in Lancasterian representation is the book. By printing large posters and occasional outsized volumes, visible by many at once, Lancaster saved on book-buying.[14] While Lancaster's system might be said to distill Western literacy practices into their most strenuous forms, emphasizing standardization, physical embodiments, repetition, routinization, and arbitrary but highly disciplined order, in this scheme these practices are not associated with the codex. In the long run, this lack may have contributed,

Figure 3.1   Students in "draughts" at "reading stations." There's a kind of mise-en-abîme sensation to this image, conveying the infinite replication not only of the system but of the students. From Joseph Lancaster's *The British System of Education,* Washington, 1812. Courtesy, American Antiquarian Society.

if in unarticulated ways, to the monitorial system's falling out fashion by the mid-century, when explicitly affective pedagogy, centered on the sentimentalized artifact of the book, would hold sway. And it signals the fact that Lancaster's system intervenes in educational history at a moment when the status of the book as the central vehicle and icon of literacy, particularly for a mass readership, is by no means ensured.

    If for Lancaster literacy is transmitted without the body of the book, so too general instruction is transmitted without the body of the teacher. But the absence of the master's person signals the ubiquity of his effect. Like a colonial administrator, this master is not subject to representation. Rather, all the powers of overview belong to him: to represent, to look, to inspect, to monitor.

> The master should be a silent by-stander and inspector. What a master says should be done; but if he teaches on this system he will find the authority is not personal, that when the pupils, as well as the schoolmaster, understand how to act and learn on this system, *the system,* not the master's vague, discretionary, uncertain judgment, will be in practice. A command will be obeyed by any boy, *because it is a command,* and the whole school will obey on the common, *known* commands of the school from being merely *known* as such, let who will give them.[15]

Such a system demands an architecture in which all the children are visible at once (figure 3.2). The floor slants, higher at the back, lower at the front, so that the master's ele-

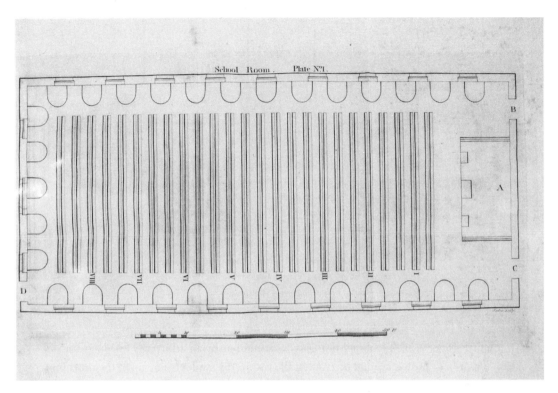

Figure 3.2   Floor plan. From the Alpha-position, the master can oversee the entire space. "The entrance door, should be on the side of the platform, at the master's end of the school, in order that the visitors on entering, may have a commanding view of all the children at once." From *Manual of the Lancasterian System, of Teaching Reading, Writing, Arithmetic, and Needle-work, as Practised in the Schools of the Free-school Society, of New-York.* New York, 1820, 6.
Courtesy, American Antiquarian Society.

vated position offers a clear line of sight to the back of the room. The children are placed in order to be observed, first for purposes of discipline: "The consciousness of being under the master's eye," writes Lancaster, "has a tendency to prevent half the usual school offences." Like nineteenth-century prisons, which were real tourist attractions, Lancaster's schools were, to a lesser degree, also sights: "This arrangement of the school has another advantage: visitors, on entering the school-room have a full view of the whole school at once: a sight to the benevolent heart, most interesting, and to the eye, one of the most pleasing which can be witnessed."[16]

Along with this panoptical visibility[17]—warming the heart as it disciplines someone else's body—Lancaster's system produces as well a certain legibility. In a classroom cryp-

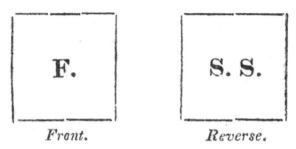

Front.   *Reverse.*

Figure 3.3   Telegraph. "The telegraph placed at the head of the
school consists of six squares, each square abour four inches by
three. These squares play on pivots, in the sides of a wooden
frame. On each side is a letter as F. *front,* on seeing which, the
whole school face the master: or, S.S. as show slates, on which the
whole school show slates. The attention of the school is called to
this by means of a very small bell *affixed,* which does not require
loud ringing, but has a sharp clear sound." From *The Lancasterian
System of Education, with Improvements* (Baltimore, 1821), 10.
Courtesy, American Antiquarian Society.

tography, students are identified not only by name but by an arbitrarily assigned "mustering number," posted around the wall of the room; the students line up in the morning
under their number, and in this way, the monitors can quickly note the truants.[18] In the
Lancasterian system, the visibility of the child does not demand the transparency promoted by evangelicism, in which the child's comportment reveals the state of its soul.[19]
Rather, these students are figured as all exterior, all surface. The time-honored pedagogical trope of the written-upon child is expressed most famously in Locke's "tabula rasa";
"impress" and "imprint" are keywords of eighteenth- and much nineteenth-century pedagogy. The Lancasterian student is emphatically not the impressible Lockean subject,
which, as developed further by Rousseau and Pestalozzi, gradually became the standard
for middle-class white schooling into the nineteenth century. Locke's is a trope that printing makes possible, while Lancaster incorporates a psychology aligned to a different communications technology. The Lancasterian student is not so much a surface to be written
on as he is a surface upon which messages can be posted.

   Among Lancaster's innovations are a number of media-technology gadgets, notably
something he calls the "telegraph" (figure 3.3). These are signs with a series of codes on
them, giving commands; in figure 3.3, "F" means to *face front* and "SS" to *show slates.* Leaving nothing to chance, Lancaster illustrates the actions that are meant to be motivated by

these commands (figure 3.4). Lancaster models his telegraph system not, of course, on the electromagnetic telegraph, which, though theorized in the eighteenth century, doesn't come into use until Samuel Morse's practical innovations in the 1830s. Lancaster would be thinking instead of the system of optical or semaphore telegraphs that spread across Europe and parts of the United States in the 1790s and early 1800s (figure 3.5). Primarily a military technology, the optical telegraph was invented in France by Claude Chappe and functioned by encoding the alphabet in arrangements of articulated arms or shutters that could be witnessed by a telescope operator at the next station, about six miles (ten kilometers) away. By 1805 the French had lines from Paris to Brest, Boulogne, Lille, Brussels, Metz, Strasbourg, and Lyon. Military and shipping news that would conventionally take days could be conveyed within hours. Between 1801 and 1807, a line was built from Martha's Vineyard to Boston and others were proposed but never built between Maine and New Orleans.[20]

Circa 1800 technologies of communication and transportation were closely allied, tied to the same networks, and those who were interested in the circulation of information—what the era called "the diffusion of knowledge" and what we would call information or data networks or media—were interested in both. The American Christopher Colles, who proposed an extensive system of optical telegraphs for the United States, was an important geographer and published the first book of U.S. roadmaps. Then as now, the desire for information goes hand in hand with the categorizing and regulating impulse. Indeed, the optical telegraph and the analogous naval semaphore became famous for transmitting two imperial dicta: "Paris est tranquille et les bons citoyens sont contents" (Paris is quiet and the good citizens are content; Napoleon, from Paris to the Provinces, in 1799); and Lord Nelson in 1805, to the fleet: "England expects every man will do his duty."[21] Both might be said to be translations into capital letters of Lancaster's classroom slogan; to put it another way, Lancaster's gift was to condense complex social regulation into statements whose proverbial ring makes them legible to the nursery crowd.

Having instituted the classroom telegraph Lancaster found himself with an invention that exceeded his original motives; his telegraph was apparently being appropriated by his instructors for their own purposes: "With some teachers the rage is a telegraph for everything, and if a telegraph could have brains or communicate intellect, too much could not be said of its importance." With the anxious if sardonic hint toward artificial intelligence, Lancaster expresses the paranoia that accompanies the valorization of media technology—a fear of the technology's seemingly self-authorized power. Lancaster lamented that his followers took to the term "telegraph" faddishly and was exasperated by those

Figure 3.4   Postures. Here boys display the correct postures in response to "S.S." (Show slates). Notice the hats on their backs, attached by string, which they toss over their shoulders when commanded to "sling hats." From *The Lancasterian System of Education, with Improvements* (Baltimore, 1821), 27. Courtesy, American Antiquarian Society.

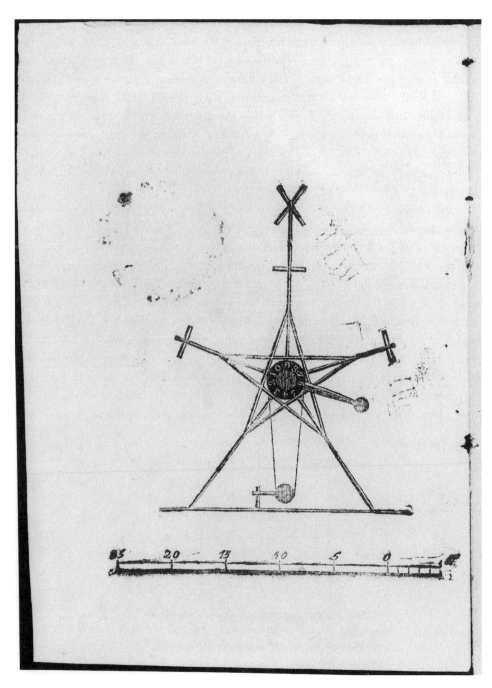

Figure 3.5   Christopher Colles's improvement on the optical telegraph, like Lancaster's "alphabet wheel,"
relies on "a revolving index" on a circular board. From *Description of the Numerical Telegraph.* Brooklyn, 1813.
Courtesy, George Peabody Library, Baltimore, Maryland.

who would insist on using the word "telegraph" to refer to other classroom technologies that, by his lights, were certainly not telegraphs: "*Examination sticks*—These answer the purpose simply of shewing when an examination has been made; some ignorant persons call them telegraphs" (figure 3.6).[22] The difference is, in fact, a crucial, if unarticulated one, between the telegraph and the "examination stick" (which, in the example here, indicates simply that an exam is going on in class group number 8). The exam stick merely conveys data, while the telegraph literally incorporates the student into a network; only when the student performs the cued action can the network proceed. The Lancasterian child is thus positioned not only inside of but as a moving part within a kind of knowledge machine.

Mainly used to communicate to a student the knowledge of his body's own movements and placement, the telegraph is part of Lancaster's general mania for spatial distortion as a means of regulation. The architectural historian Dell Upton has described the Lancasterian system as one which conceived "of human knowledge, orderly instruction, individual accomplishment, and spatial order as interlocking realms in which the material world could act as a tool for shaping the moral and intellectual faculties decisively and efficiently."[23] Lancaster manipulates these "interlocking realms" by shifting scale; he loves to expand, to contract, and in effect to create parodic, even gothicized, versions of ordinary human scale in a pedagogical *Castle of Otronto*. In his classroom, not fifteen or fifty, but five hundred students in one room. Not just illustrations in his books, but elaborate foldouts. Not ordinary schoolbooks, in the hands of the children, but outsized versions to be viewed from afar. Not ordinary speech, but sequences of coded commands, requiring

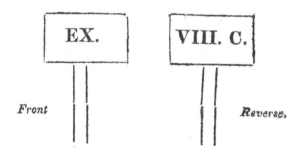

Figure 3.6   Examination sticks. (Not to be confused with telegraphs.) From *The Lancasterian System of Education, with Improvements* (Baltimore, 1821), 3. Courtesy, American Antiquarian Society.

specialized knowledge. These protocinematic effects—of zoom, close-up, long-shot, and tracking—displace identity, keeping the child off balance, so that he can only find his place within the system itself. I use the analogy to cinema advisedly, for Lancaster is attuned to the medial shifts of his own epoch, as expressed by his passion for the telegraph and disdain for books (other than his own). Shifts in scale like the ones he institutes regularly occur with shifts in media technology: The distant is brought close; the small is made large; the prolix is condensed.

That which in the realm of literature for children, beginning in the eighteenth century, would be the miniature, becomes, in Lancaster's universe, the abbreviated, the, literally, telegraphed. While both modes signal a consensus that material for children ought to be scaled down and condensed, their differences mark the boundaries of an emerging normative view of childhood, which excludes the poor and children of color. The middle-class child, the object of the boom in children's publishing, becomes a vehicle for adult memories and desires; as such he is (if often ambivalently) encouraged in a certain kind of sensuous, internalized reading experience. The story of Tom Thumb, toy books, tiny bibles no larger than postage stamps—when adults market such objects to children they tap a realm that enacts a "bourgeois longing for inner experience." The miniature, Susan Stewart writes, inspires an "infinite time of reverie" and creates an "'other' time, a type of transcendent time which negates change and the flux of lived reality."[24] By contrast, the abbreviation is profoundly sublunary; far from slowing time down, the abbreviation willfully accelerates it. If the miniature allows for a kind of narcissistic flood, the abbreviation is grounded in the social. It even creates its own class around the literacy event that it inspires: You're "in the know" if you find the abbreviation legible. In this way the abbreviation embraces change; it displaces the anxiety inspired by modernity's plunge into futurity onto the very act of reading, transforming incompetency and inexperience into a sense of mastery, of knowingness, and of belonging.

But at the same time that the telegraph and its methods constitute an accommodation to modernity—this is, after all, part of what made Lancaster's system popular—they also create a temporal and spatial density from which it is hard to remove oneself: no time for thought or privacy or interiority. These are for others. David Wallace Adams describes the boarding school experience of an Indian student, at a much later date, which neatly captures the Lancasterian ethic: "Every aspect of his day-to-day existence . . . would be rigidly scheduled, the hours of the day intermittently punctuated by a seemingly endless number of bugles and bells demanding this or that response."[25] Lancaster, too, like so many of his pedagogical heirs, was devoted to bells, which punctuated the day, and signaled when a "telegraph" was arriving.

Telegraph: literally, writing at a distance, from afar. But why do you need a telegraph when you are in the same room? When is proximity figured as distance? One answer is that hierarchy always safeguards class, age, gender, or racial distinctions through protocols of communication; but in this case, something more is at work. During the formation of the republic, education was conceived of as a component of the "diffusion of knowledge," essential to the transformation of "subjects" into "citizens." Such "diffusion" implies an existing cache of "knowledge" requiring radiation from a center along metropolis-to-provinces routes. Aligned with this mapping, the Lancasterian structure positions "monitors" as relays between master and pupils; in the classroom telegraph, this relationship is distilled and refined into ever more narrow channels.[26] The paradigm for education, formerly the guild and scholastic model of master-apprentice, of vertical ascent in a "tower of learning" or through degrees of mastery, is here represented as a horizontal network, where face-to-face interaction is increasingly reduced, authority is increasingly abstracted—and facelessly enforced—in a model suggested by the new media technology of the telegraph, supplanting both the rhetorical and print-based models.

The abridgement of face-to-face communication in the monitorial system transforms punishment as well; while pedagogical norms of the time still support corporal punishment, Lancaster's discipline is, by contrast, discursive and narrative. Lancaster describes his notions of disorder and the punishments designed to prevent it in distinctly orientalist terms: "The attempt to promote learning without the principle of order, would be like the efforts of the eastern nations, when Nimrod, in the despotism and pride with which he built the *Tower of Babel,* only succeeded in producing confusion, and thereby founded the first *empire of ignorance.*" Here authority, linked with the telegraph as one of the means by which authority remains consistent and anonymous, is constituted through opposition to the East, to the Orient.[27]

Lancaster's orientalism extends to his classroom literacy practices. Alphabetic technology in Lancaster's classroom reproduces an imagined history of alphabetic writing: the youngest children, the "alphabet boys," work at the "sand-table," a black sand-covered surface in which students would write with their fingers. Sand writing was an innovation of Andrew Bell, who reputedly learned it in India, by observing how the native children learned to write. Lancaster positioned the sand tables within the factory of the schoolroom, as the first step in a process. Once the letters were thus incorporated grittily into the body, the students were ready to respond to more advanced classroom technologies, including the telegraph and an "alphabet wheel," echoing Colles's telegraph (figures 3.7 and 3.8).

The punishments that Lancaster substitutes for corporal discipline similarly rely on orientalist theatrics along with shaming written labels and are brought into play when an accompanying system of rewards and merits fails. While Lancaster fosters competition, mirroring the economy, with students continually changing places, his disciplines more rigidly place students. "Birds in a cage" are placed in baskets suspended from the rafters. Often the placing is figurative and explicitly theatrical:

> instead of recurring to the rod, make him *a bashaw of three tails.* The use of the famous coat, called the fool's coat, is well known in schools; let such a coat be suspended in public schools, the name of the offender printed in large letters, that the whole school may read, and fasten on it the words "Bashaw of three tails," also on the back of the coat, and three birchen rods suspended from the tail of the coat, at due and regular distances. This punishment is excellent for the senior boys, and will not need many repetitions.[28]

A mainstay of orientalist fiction and theater of the eighteenth and nineteenth century, the bashaw is a role with antic possibilities; but for the offending boy, the performance is framed by text, fixing it within the realm of regulation rather than misrule.[29]

Lancaster further deploys an urban marketplace effigy to penalize through ridicule: "When a boy gets into a singing tone in reading . . . decorate the offender with matches, ballads, &c. and, in this garb, send him round the school, with some boys before him, crying 'matches,' &c. exactly imitating the dismal tones with which such things are hawked about the streets in London." Sean Shesgreen points out that the crier of matches is "the poorest and most numerous of street sellers."[30] Both the Nimrodlike bashaw and the abject hawker are figures of marketplace carnival; what the two have in common to the pedagogue's mind is their festive and oppositional relation to western literacy and to the technologies that transmit it. And both suggest that the student metamorphoses into a racialized "other" when he falls out of the network's order.

The monitorial classroom is, perhaps above all, a zone of mimicry;[31] indeed "emulation" ("to copy or imitate with the object of equaling or excelling," according to the OED) is an important keyword of the system. These disciplinary effigies are a variation on the theme of emulation. Given their nature, one wonders how Lancaster's system could be translated to the project of educating children of color. And yet, the system was virtually designed to teach the sons of London's street hawkers, and to wean them from the streets. While the punishments are included in Lancaster's first U.S. manual (1812), and were therefore part of the training of missionaries who were sent into the field in the teens, the

Plate 3.

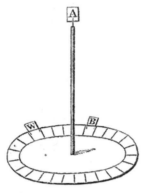

Moveable Stand.

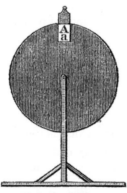

Alphabet-Wheel.

Bench with holes for Hats.
2

Figure 3.7   The Alphabet Wheel. Like the telegraph,
this anticipates later inscription technologies—for
example, the typewriter. From *Manual of the Lancasterian
System, of Teaching Reading, Writing, Arithmetic, and Needle-
work, as Practised in the Schools of the Free-school Society, of
New-York.* New York, 1820. Courtesy, American
Antiquarian Society.

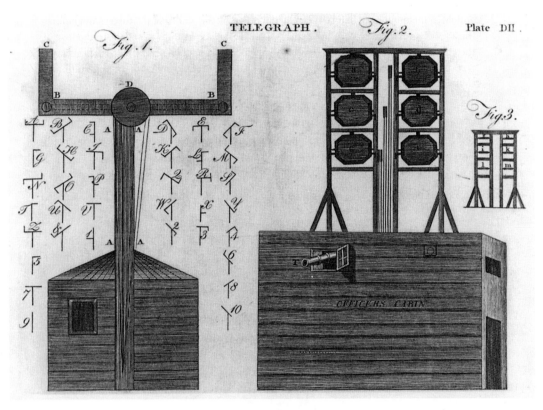

Figure 3.8   The optical telegraph. From *Encyclopedia Brittannica* (Philadelphia, 1798). Courtesy University of Minnesota Library.

punishments were dropped from the 1821 pamphlet (likely to have been circulated to mission schools). In what follows, I turn to the case of the Brainerd School in the teens and twenties, to consider some of the ways in which this pedagogy, and particularly its replicating and representational techniques, was adapted to suit the mission project.[32]

A missionary/colonialist streak runs through Lancaster's life and work, from his first impulse to run away to Jamaica (figure 3.9). He joined the Quakers not only because they could supply him with financial support, but because of their abolitionist and missionary work. When Lancaster got to the United States, he promoted his system precisely to teach American Indians, and proposed to President Monroe that monitorial methods would enlighten the mind's eye before it can be darkened by the gloom of the forest, and before the affections of the heart can be absorbed, by the furious feelings attendant to the love of the chase. Thus you will grasp a new power, which may operate on the minds of this interesting people, and render the yet unvitiated and undirected energies of their youth, subservient to the promotion of the general welfare.[33]

For the purposes of schooling, missionaries and pedagogues, like later ethnographers, conflated Indians, Africans, and the poor into an undifferentiated population, requiring the supplement of "civilization" (an eighteenth-century neologism) to render them fully human.[34] While each case—of the black, the poor white, the Indian—had its urgent claim to benevolence, the rhetoric surrounding the Indian juxtaposed, or opposed, to

Figure 3.9   This broadside image from 1813 captures the imperialism of the mission project. Here King George hands the Bible to a generic barefoot peasant or indigenous child. "Royal British System of Education." Courtesy, American Antiquarian Society.

"civilization" the stringent "extermination" and "extinction" and "extinguish." While the spatial euphemism "removal" would come into play as a way to defer both "civilization" and "extermination," there was, strikingly, no other middle term. This situation created a particular emergency for the mission project.

Confronted with the enormity of the problem of education, theorists tended to shiver under the weight of a kind of mathematical sublime. Sending missionaries off to teach American Indians in 1815 (in the Lancasterian system), the Reverend Elijah Parrish worried that

> there may be three million savages on this continent, perishing for the bread of life. Here then we have in our own country, as it were in our own neighborhood, in our own family, six million souls, sinking into ruin, crying to us for help, to pluck them as brands from the burning. Is not here a field of spiritual misery and death, far surpassing all our resources, far surpassing the labors of all our missionaries? Six million souls need six thousand missionaries. But where can half, where can one tenth part of this number be found?[35]

The occasion for this sermon was the ordination of Cyrus Kingsbury, who was being sent to the Cherokees by the American Board of Commissioners for Foreign Missions, the Massachusetts congregationalist organization responsible, from their founding in 1810, for the largest number of missions to American Indians. By the time the mission school at Chickamauga, Tennessee, was opened in 1817, the Cherokees were in extremis, riven by internal pressures for and against assimilation and by external pressures exerted by the whites surrounding them. As Margaret Szasz has pointed out, the "success" of mission projects depended largely on "the extent of disruption experienced by [the] native culture."[36] By this measure, Kingsbury's school would be destined for great things.

On a mission tour in 1818, the American Board's treasurer, Jeremiah Evarts, wrote from Kingsbury's newly christened Brainerd station that the Lancaster plan is "not only excellent in itself, but particularly suited to the attention of Indian children."[37] This assertion goes unelaborated, but in a later account of Brainerd, the narrator relates the visit of a Cherokee chief to the classroom: "The king [chief] addressed the scholars in Cherokee. . . . It was ascertained from the bi-lingual children that the king's sentiments corresponded exactly with his gestures, so that the missionaries understood what he uttered, by his gesticulations, almost as well as if he had made an address in correct English."[38] This widespread nineteenth-century notion of the gestural language of Indians, the idea that Indians can thus be fully and accurately read from the outside, without the wearisome bother of translation, is part of what made the Lancasterian system seem plausible to missionaries and policy makers.

At the same time that Indians were regularly drawn into English literacy as a constituent of the civilizing process, Anglo-Americans and Europeans consistently noted what they perceived as a special native gift for oratory. This characteristic veneration for Indian oratory has an elegiac cast. David Murray argues that here, as in much American discourse about Indians, the notion of Indian "nobility" aligns the Indian with the moribund aristocracies of Europe. Further, Murray points out, Indian speech, when not perceived as prelinguistic altogether, was essentialized as authentic nature; he cites an 1836 article on "Indian Eloquence" that remarks that the "'genius of eloquence bursts the swathing bands of custom, and the Indian stands forth accessible, natural, and legible.'"[39] It is in this familiar "legibility" that this imagined Indian might be easily positioned within Lancaster's network. As in the case of the Lancasterian child noted above, the Indian is represented as lacking a unique interiority—a self or subjectivity—and therefore might be easily provided with a surrogate self—or, more precisely, a *representation* of a self—in its stead.

The Brainerd school at Chickamauga used the Lancasterian devices of monitors and merits, including gifts and money. Monitors functioned to relieve the small missionary staff, who were also responsible for every aspect of life for their families and the Indian students who boarded with them. Lancaster had a Wizard of Oz–like set of medals that he gave out to students, which attracted the waggery of his critics; in America these rewards, which began as leather or metal badges and medals, were supplemented and replaced by script, tickets, and tokens exchangeable for "such articles as the holders need" and which "answer the purpose of a circulating medium among the boys."[40] Combining rewards with codes echoing the telegraph system, Cyrus Kingsbury distributed "little cards bearing the initial letters of the words 'Punctual attendance,' 'Behavior,' and 'Diligence.'" These coded cards could then be redeemed for cash: one-half cent, one cent, and three-and-a-half cents respectively. The cash could be exchanged for books, knives, or toys.[41] Cherokees had long been incorporated into the white economy, but this system associates literacy and the high-rent "diligence" (a virtue much urged upon Indians who were thought to be "idle" and "indolent") with the consumption of goods.

It was not only in such commodity practices that Indians were brought into the white economy. Through the naming of both mission schools and mission students, Indians were incorporated into white culture through a kind of sentimental cryptography, in a code legible only to the whites. Brainerd, Dwight, Eliot—these American Board mission schools, to the Cherokees, Choctaws, and Creeks, were named to commemorate and perpetuate the work of the previous generations of missionaries and teachers: John Eliot (1604–1690) to the Massachusett; David Brainerd (1718–1747) to the Seneca and Delaware; and the educator and minister Timothy Dwight (1752–1817). Much is made in the nineteenth century of Indian place names. "How can the red man be forgotten," writes the poet Lydia Huntley Sigourney, "while so many of our states and territories, bays, lakes and rivers, are indelibly stamped by names of their giving?" The carrying of the New England names into the Southeast reverses this process, imprinting the landscape with these cenotaph inscriptions.

Such names are replicated as well when the schools rename their Cherokee students after dead missionaries and living benefactors, incorporating them into an institutional rather than a familial kinship structure.[42] Thus "Brainerd" becomes not only the name of the mission school, but the name of one of the Cherokee students—"David Brainerd"— in the school. This renaming places students within a new constellation, legible only from the outside; like little Bellerophons, they are, at first anyway, carrying a language that they can't read into a culture that wishes them ill.[43]

In keeping with Lancasterian accounting procedures, Brainerd teachers tracked each student in their record books, listing the child's name and age, when he or she came to the school, his or her Cherokee name and its English translation, and in the "Character" column such attributes as "Respectable," "Pious," "Industrious"; "Mediocrity" or "Saucy"; and, once, "a wild creature just from the woods." With a double-entry attention to the result of their investments, some records concluded with "Runaway" or "Died." The list of names reads like entries in the *Dictionary of American Biography,* dizzyingly replicating the American Board's theological, political, and literary taste, as though the students were to form a living canon. From the *Brainerd Journal*'s tables: Samuel Worcester (twice); John Knox Witherspoon ("Run away"); Thomas L. McKenney ("Not so attentive as he should be"); Lydia Huntley; Boston Recorder; Lyman Beecher; Timothy Dwight; Jedidiah Morse; Jeremiah Evarts; Elias Boudinot.[44] The names ring a change on Lancaster's classroom cryptography described above. The poor are identified by number, but in a sentimentalizing of Lancaster's militaristic and bureaucratic "mustering," the Indians are identified with names richly enmeshed in the web of United States culture: its literacy practices, its memorializing customs, its affective demands, its patrilineal kinship structure, its institutional filiopiety, its processes of commodification.

The renaming serves several practical purposes:

> In compliance with the request of a society of young gentlemen in Southampton, Mass. a boy has been selected to bear the name of Vinson Gould, to be educated by them & at the request of a society of young ladies in the same place a girl has been selected to bear the name of Mindwell Woodbridge Gould, to be educated by them. The children are called after their pastor & his wife. . . . [The little girl] being destitute of a name that we could conveniently pronounce . . . has been called baby till the present time.[45]

Here the names make good on a debt, memorialize a "pastor & his wife," and resolve the continually vexing problem of many missionaries' monolingualism. (Presumably "Mindwell Woodbridge Gould" would rarely have to pronounce her own name.) To affix a name was to fix the Indian within a new social space. The names also functioned as apotropaic charms to prevent the Indian from engaging in the kind of movement represented by Cherokee names, which often have an active and somewhat narrative quality.[46]

What did the students make of their new names? Ethnographers suggest that for Cherokees, a name was not a kinship marker, but was meant to carry particular information about the individual; it had an active, not a memorializing or retrospective relation to the person, and was capable of being changed as the events of life unfolded, demanding a new name.[47] For most native Americans, a "degree of sensitivity to the use of the name or names of dead persons was so common that its absence is noted only a few times in the ethnographic literature. . . . Respect and consideration for the bereaved and fear of the dead entered into . . . restrictions" on being named for the dead.[48]

With their missionary necronyms, the students became living monuments. At the same time, they also became promotional devices; their letters to their benefactors were advertisements for Indian progress. As the children became fluent, literacy was put to work in the interests of warding off further harm. Cherokees and those in sympathy with them hoped that removal from their established homes and ancestral land to unknown western territories could be forestalled by accumulating sufficient evidence of Cherokee civilization. To this end, children at Brainerd in the 1820s routinely wrote to their benefactors. On January 1, 1829, Nancy Reece wrote to Reverend John Johnston: "I shall be pleased to hear from New Burg as we have named a little girl Mary Johnston I hope you will write often." And on June 23, 1828, Lucy Campbell closed a letter to Daniel Campbell "from your adopted Cherokee daughter."[49] In addition to reminding their correspondents of their financial commitment, these letters formulaically refer to what white consensus of the period regarded as particular markers of civilization: the looms and spinning wheels of the women; the hoes and ploughs of the men; and the reading and writing of the children.[50] While these objects are advertisements for the Cherokee's becoming incorporated into an economy of consumption, they, like the renaming, signal that the Indians are becoming fixed in place.

There is a great deal more to be said about Nancy Reece and her student colleagues; their letters are eloquent in themselves and are striking examples of the literature of childhood as well as of Cherokee literature. Even given the difficult circumstances of their composition, just preremoval during the Jacksonian transformation of Indian policy, these letters convey their writers' unique voices, as well as expressions of political will, as when the children write directly to Andrew Jackson. In some ways, these letters seem far removed from Lancaster's original system.

And yet, whatever else the letters may do, they position the students as vehicles of white memory and as publicity agents for the missionary project of which they themselves are the object. Here the order of "civilization" is instituted by an inversion of the disciplinary effigies noted above; these Indian children are written into a masquerade in which they function as monuments and celebrations, as replicas and repetitions. It is in this representational work that the letters display their consonance with the monitorial project, in which the child is produced as, in effect, a medium of communication.

The relationship between media technologies and the socialization of children in the nineteenth century is often described in terms of the valorization of books and reading and the expansion of print culture through mass literacy. In contemporary culture the discussion about media technology (particularly the internet) in relation to children often focuses on access to the equipment, on one hand, and control of the content on the other. The example of Lancaster's engagement with new media at the turn of the nineteenth century offers a site for thinking about other ways in which media technologies interact with and impose upon children, in both theory and practice. The Lancasterian system was discredited by theorists of the mid-century precisely for its mechanistic innovations. But the system's effects would be long-lasting, for by then its representational techniques, its miming of technologies of communication, and its production of technologies of the self had become inextricably embedded in urban schooling and mission schooling—sites, that is, in which students' interiority was intentionally proscribed.

Acknowledgments

I wish to acknowledge the American Antiquarian Society, the American Society for Eighteenth-Century Studies, and the Newberry Library-Spencer Foundation Fellowship in the History of Education for the funding of research for this essay.

Notes

1. Joy Harjo and Gloria Bird, *Reinventing the Enemy's Language: Contemporary Native Women's Writings of North America* (New York: Norton, 1997), 469.

2. David Wallace Adams, *Education for Extinction: American Indians and the Boarding School Experience 1875–1928* (Lawrence: University Press of Kansas, 1995), 120.

3. A sketch of the phrase's lineage shows how specific to the bureaucratic sensibility and how modern Lancaster's meaning is. The OED misses the 1812 Lancaster source: Joseph Lancaster, *The British System of Education* (Washington, 1812), 2. Instead the OED cites first use in Marryat's 1842 *Masterman Ready:* "In a well-conducted man-of-war . . . every thing is in its place, and there is a place for every thing." Isabella Beeton's 1861 *Book of Household Management* uses the expression in much the same way (OED). Two related dicta emerge from antiquity: Horace on stylistic decorum (*Singula quaeque locum teneant sortita decentum,* let each [style] keep the becoming place allotted it [De Arte Poetica, I. 92]) and Augustine, like Horace relying on rhetorical tradition: "Order is an arrangement of components . . . assigning the proper place to each" (*City of God,* 19, 13). In his 1642 book of aphorisms, *Jacula Prudentum,* George Herbert lades the idea with spiritual anxiety on the one hand—"All things have their place, knew wee how to place them"—and status anxiety on the other: "Sit in your place, and none can make you rise," the latter a version of the traditional "He need not fear to be chidden that sits where he is bidden." Benjamin Franklin neatly combines the notion of the decorous style with a tip on general deportment: "Tim and his Handsaw are good in their Place,/Tho' not fit for preaching or shaving a face"("Poor Richard," 1746). See John Chapin, ed., *Book of Catholic Quotations* (New York: Farrar Straus and Giroux, 1956); Meider, Wolfgang, et al., *Dictionary of American Proverbs* (New York: Oxford University Press, 1992); *Oxford Dictionary of English Proverbs,* 3d ed. (Oxford: Clarendon Press, 1970); and Stevenson, Burton, *The Home Book of Proverbs, Maxims, and Familiar Phrases* (New York: Macmillan, 1948).

4. Peter Hulme defines "colonial discourse" as "an ensemble of linguistically based practices unified by their common deployment in the management of colonial relationships. . . . Underlying the idea of colonial discourse . . . is the presumption that during the colonial period large parts of the non-European world were *produced* for Europe through a discourse that imbricated sets of questions and assumptions, methods of procedure and analysis, and kinds of writing and imagery, normally separated out into the discrete areas of military strategy, political order, social reform, imaginative literature, personal memoir and so on"; *Colonial Encounters: Europe and the Native Caribbean, 1492–1797* (New York: Methuen, 1986), 2.

5. National Archive, Bureau of Indian Affairs, War Department, Secretary's Office, Letters Sent, Indian Affairs, E: 151. In a letter dated August 29, 1821, and initialed by Secretary of War John Calhoun's chief clerk, Christopher Vandeventer: "By direction of the Secretary of War, I transmit to you a pamphlet relative to the Lancasterian System of Education, by its founder. This system it appears has been adopted, with advantage, in some of the Indian schools already in operation, and the propriety of adopting it generally is suggested." Robert Berkhofer very plausibly suggests that the pamphlet is Lancaster's 1821 Baltimore manual, *The Lancasterian System of Education, with Improvements* (see *Salvation and the Savage: an Analysis of Protestant Missions and American Indian Response, 1787–1862,* Lexington: University of Kentucky Press, 1965, 27n39).

6. This anecdote is related in David Salmon, *Joseph Lancaster* (London: Longmans, Green, 1904), 2–3. The Clarkson essay would have been either his prize dissertation at Cambridge, *An essay on the slavery and commerce of the human species, particularly the African* (1786) or *An essay on the impolicy of the African slave trade* (1788). Clarkson was a prominent Quaker abolitionist; one of the Pennsylvania Abolition Society's Schools, which also adopted Lancaster's plan, was called the Clarkson school.

7. Bentham's *Chrestomathia* (1816) draws heavily on Lancaster as well as on Andrew Bell.

8. Carl Kaestle, *Joseph Lancaster and the Monitorial School Movement: A Documentary History* (New York: Teachers College Press, 1973), 40. It is his nonsectarianism as well as his self-promoting personality and the visuality of his manuals that made Lancaster a draw in America. Lancaster drew some aspects of his program from Andrew Bell's Madras system, which, like Lancaster's, relied on and participated in colonial discourses. Sources for Lancaster's biography include David Salmon, *Joseph Lancaster;* John Franklin Reigart, *The Lancasterian System of Instruction in the Schools of New York City* (New York: Teachers College, 1916); Lancaster's own *Epitome of Some of the Chief Events and Transactions in the Life of Joseph Lancaster, containing an account of the rise and progress of the Lancasterian system of education; and the Author's Future prospects of usefulness to mankind* (New Haven, 1833); Mora Dickson's *Teacher Extraordinary Joseph Lancaster 1778–1838* (Sussex: The Book Guild, 1986); and Kaestle's anthology.

9. Carl Kaestle notes that the English solution to threat of unrest from below was to educate the poor only within strict limits, meant to impress upon them their place. In the early republic, the poor, like the native population, were deemed capable of "civilization," and white middle-class culture was seen to benefit from this; Kaestle, *Lancaster,* 2; and Carl Kaestle, *Pillars of the Republic: Common Schools and American Society, 1780–1860* (New York: Hill and Wang, 1983), 33–37.

10. Robert H. Bremner, ed., *Children and Youth in America: A Documentary History* (Cambridge: Harvard University Press, 1970), viii; Jacqueline Reinier, *From Virtue to Character: American Childhood 1775–1850* (New York: Twayne, 1996), 50.

11. For Lancaster and the market economy, see David Hogan, "The Market Revolution and Disciplinary Power: Joseph Lancaster and the Psychology of the Early Classroom System." *History of Education Quarterly* 29, no. 3 (fall 1989): 381–417; for Lancaster and architecture, see Dell

Upton, "Another City: The Urban Cultural Landscape in the Early Republic." In *Everyday Life in the Early Republic,* ed. Catherine E. Hutchins (Winterthur: Winterthur, 1994.)

12. Kaestle, *Lancaster,* 14, passim.

13. See David Hogan, who describes Lancaster's pedagogy as "a manufactory of desire and ambition" (384). See Laura Rigal on manufactory in the early republic in relation to representational technologies, *The American Manufactory: Art, Labor, and the World of Things in the Early Republic* (Princeton: Princeton University Press, 1998), 13–17. System, notes Kaestle, "was the essence of technology: it was not simply 'efficient,' . . . but . . . it was infinitely replicable" and this "idea of a uniform, recursive system is consonant with the developing technological ideology of the nineteenth century" (*Lancaster,* 15).

14. At his school in Southwark, London he had his own printing press, on which he produced some of these texts. Bound together in one large (17″ × 21″) volume, for example, are the *New Invented Spelling Book* and *Freames Scripture Instruction,* along with broadside ephemera—Samuel Whitchurch's poems "My Mother" and "The Bible." While the catechism and poems are conventional, the spelling book reveals Lancaster's eccentric methodology, in which the letters are taught out of order, arranged instead according to their visual similarity.

15. Lancaster, *British System,* 93; emphases in original.

16. Joseph Lancaster, *Hints and Directions for Building, Fitting Up, and Arranging School Rooms on the British System of Education* (London, 1809), 11, 13; see John Bender for tourism and the eighteenth-century British prison; *Imagining the Penitentiary: Fiction and the Architecture of Mid-Eighteenth-Century England* (Chicago: University of Chicago Press, 1987), 14.

Prison reform and school reform tended to be inaugurated by citizen committees and government agencies on whose rolls the same names turn up again and again. The rhetoric of schooling in the early republic always adverts to the costly alternative, the prison; an anonymous letter about Lancaster in the *Niles' Register* April 24, 1819 (150) is one of numberless examples: "To *enlighten the mind,* and thereby *prevent crime,* is better than to punish or commiserate." Joined here in the figure of chiasmus, the two institutions generally share benefactors and theorists as well. In the "Report . . . to the Lancaster School Society of Georgetown" in 1812, the trustees noted that "out of six thousand persons educated by [Lancaster] alone in one school [in London], there has never yet occurred an instance of any one of them being called before a court of justice on a criminal accusation" (Lancaster, *British System,* 123).

17. "Hence the major effect of the Panopticon: to induce in the inmate a state of conscious and permanent visibility that assures the automatic functioning of power. So to arrange things that the surveillance is permanent in its effects, even it if is discontinuous in its action . . . that this architectural apparatus should be a machine for creating and sustaining a power relation independent of the person who exercises it"; Michel Foucault, *Discipline and Punish: The Birth of the Prison,* trans. Alan Sheridan [1977] (New York: Vintage Books, 1995), 201. Foucault is speaking here of Bentham's carceral system, a plan closely allied to pedagogical schemes, including Lancaster's.

18. Joseph Lancaster, *The Lancastrian System of Education, with Improvements* (Baltimore 1821), 3. Dickens's *Hard Times* [1854] (New York: Bantam Books, 1981) is apparently about a Lancasterian school: Sissy Jupe is also known as "girl number twenty" and students "sat on the face of the inclined plane" (3). Consonant with the sentiments of the epigraphs to this essay, Dickens figures the teacher, Gradgrind, as an embodiment of the square: "The speaker's square forefinger emphasized his observations. . . . The emphasis was helped by the speaker's square wall of a forehead . . . square coat, square legs, square shoulders" (1).

19. The notion of transparency is rooted in Puritan "visible sanctity," but evolves across the eighteenth and into the nineteenth century into what Karen Haltunnen calls "sentimental transparency"; *Confidence Men and Painted Women : A Study of Middle-Class Culture in America, 1830–1870* (New Haven: Yale University Press, 1996), 57–59.

20. For the history of the optical telegraph, see Gerard J. Holzmann and Bjorn Pershon, *The Early History of Data Networks* (Los Alamitos, Calif.: IEEE Computer Society Press, 1995); see *Encyclopedia Britannica* (1797), sv. Telegraph. See also James W. Carey, especially "Technology and Ideology," which calls for a thorough and culturally grounded history of the telegraph; *Communication as Culture: Essays on Media and Society* (New York: Routledge, 1989).

21. Holzmann, 71; OED, sv "telegraph" (citing A. Duncan's *Nelson*, 1806, 297).

22. Reigart *Lancastrian System*, 10, 3.

23. Upton, "Another City," 84.

24. This is James Clifford, "On Collecting Art and Culture," 49–73, *The Cultural Studies Reader*, ed. Simon During (New York: Routledge, 1993), 53, and Susan Stewart, *On Longing : Narratives of the Miniature, the Gigantic, the Souvenir, the Collection* (Chapel Hill: Duke University Press, 1993), 65.

25. David Wallace Adams, *Education for Extinction: American Indians and the Boarding School Experience 1875–1928* (Lawrence: University Press of Kansas, 1995), 117.

26. This figurative usage of "diffusion" dates from the mid-eighteenth century; the OED cites Dr. Johnson (1750) and David Hume (1777) among its earliest citations. "Monitor" originally means "to advise"; its subsequent evolution as a broadcast and electronic media device—substituting for the human—is worth noting.

27. Reigart, *Lancastrian System*, 10; emphasis in original. Edward Said notes that the project of analyzing orientalism shows "that European culture gained in strength and identity by setting itself off against the Orient as a sort of surrogate and even underground self"; *Orientalism* (New York: Pantheon, 1978), 3.

28. Lancaster, *British System*, 68–69; 75; emphasis in original.

29. An illustration of a "bashaw of three tails" appears in the 1829 children's game *Aldiborontiphoskiphorniostikos* by R. Stennett.

30. Lancaster, *British System,* 74; Sean Shesgreen, *The Criers and Hawkers of London: Engravings and Drawings by Marcellus Laroon* (Aldershot, U.K.: Scolar Press, 1990), 80.

31. Homi K. Bhabha uses the figure of mimicry to describe colonialism itself: "If colonialism takes power in the name of history, it repeatedly exercises its authority through the figures of farce. . . . The civilizing mission . . . often produces a text rich in the traditions of *trompe-l'oeil,* irony, mimicry and repetition. In this comic turn from the high ideals of the colonial imagination to its low mimetic literary effects mimicry emerges as one of the most elusive and effective strategies of colonial power and knowledge"; *The Location of Culture* (New York: Routledge, 1994), 85.

32. Ronald Rayman assesses what he sees as the failure of the Lancasterian system at the Brainerd School and elsewhere in the Southeast, providing a useful overview of the mission project. He sees the missionary emphasis on "useful work" rather than education as contributing to Cherokee resistance to the mission schools; "Joseph Lancaster's Monitorial System of Instruction and American Indian Education, 1815–1838," *History of Education Quarterly* 21 (1987): 395–409. See also Robert Sparks Walker's *Torchlights to the Cherokee* (New York: Macmillan, 1931) for a further account of the Brainerd School.

33. Joseph Lancaster, *Letters on National Subjects, Auxiliary Education, and Scientific Knowledge; Address to Burwell Bassett, Late a Member of the House of Representatives; Henry Clay, Speaker of the House of Rep, and James Monroe, President of the United Sates by Joseph Lancaster* (Washington City, 1820), 42.

34. "[E]thnographers tended to equate Indians to impoverished people in their own societies. Alice Kehoe has described the consequences of this approach for women: 'The traditional [ethnographic] picture of the Plains Indian woman is really that of an Irish housemaid of the late Victorian era clothed in a buckskin dress'"; Thea Perdue, *Cherokee Women: Gender and Culture Change, 1700– 1835* (Lincoln: University of Nebraska Press, 1998), 5. Locke's perception of Indians' failure to improve their property through labor is one of the origins of this error; see William Cronon, *Changes in the Land: Indians, Colonists and the Ecology of New England* (New York: Hill and Wang, 1983), 79–80. In any case it pervades white writing about Indians, who are consistently described by the American Board as "destitute," a term that combines economic and spiritual poverty. On "civilization," Raymond Williams cites Boswell's account of Johnson's disdain for the term: "'He would not admit *civilization,* but only *civility.* With great deference to him, I thought *civilization* . . . better in the sense opposed to *barbarity,* than *civility.*" Williams comments that "Boswell correctly identified the main use that was coming through, which emphasized not so much a process as a state of social order and refinement, especially in conscious historical or cultural contrast with *barbarism.* Civilization appeared in Ash's dictionary of 1775, to indicate both the state and the process"; Raymond Williams, *Keywords: A Vocabulary of Culture and Society* (New York: Oxford University Press, 1985).

35. Elijah Parrish, *Sermon Preached at Ipswich, Sept. 29, 1815. At the Ordination of Reverand Daniel Smith and Cyrus Kingsbury as Missionaries to the West* (Newburyport: 1815), 11–12.

36. Margaret Szasz, *Indian Education in the American Colonies, 1607–1783* (Albuquerque: University of New Mexico Press, 1988), 259. For the history of Cherokees, I have drawn on William G.

McLoughlin's *Cherokee Renascence in the New Republic* (Princeton: Princeton University Press, 1986); Walter H. Conser, Jr., ed., *The Cherokees and Christianity, 1794–1870: Essays on Acculturation and Cultural Persistence* (Athens: University of Georgia Press, 1994); and *Cherokees and Missionaries 1789–1839* (New Haven: Yale University Press, 1984); as well as Theda Perdue's *Cherokee Women;* and James Mooney's *Historical Sketch of the Cherokee* [Bureau of American Ethnology, 1898] (Chicago: Aldine, 1975).

37. American Board of Commissioners for Foreign Missions, *First Ten Annual Reports of the American Board of Commissioners for Foreign Missions 1810–1820* (Boston: 1834), 193.

38. Sarah Tuttle, *Letters and Conversations on the Cherokee Mission* (Boston: Massachusetts Sabbath School Union, 1830), 74; Tuttle seems to be drawing on American Board reports. This anecdote is reported as well in the American Board report for 1819 (American Board, 238).

39. David Murray, *Forked Tongues: Speech, Writing and Representation in North American Indian Texts* (Bloomington: Indiana University Press, 1991), 35–36, 41.

40. American Board, *First Ten Annual Reports,* 192.

41. Robert Berkhofer, *Salvation and the Savage: an Analysis of Protestant Missions and American Indian Response, 1787–1862* (Lexington: University of Kentucky Press, 1965), 26.

42. For Sigourney (above) see *American Poetry: The Nineteenth Century,* vol. I. (New York: Library of America, 1993), 111. The renaming constitutes part of what the sociologist Pierre Bourdieu calls a "rite of institution"; *Language and Symbolic Power,* translated, Gino Raymon and Mathew Adamson (Cambridge: Polity, 1991), 117–126. He is extending van Gennep's notion of the rite of passage to institutional performances. See also Bourdieu on naming; *Outline of a Theory of Practice,* trans. Richard Nice (Cambridge: Cambridge University Press, 1977), 36. The rite of institution is an "act of communication" that "*signifies* to someone what his identity is" (121). In most Indian school settings these missionary rites are explicitly violent; typical is a description of a Shawnee school in 1850: "The service to a new pupil was to trim his hair closely; then with soap and water, to give him or her the first lesson on godliness, which was a good scrubbing, and a little red precipitate on the scalp to supplement the use of a fine-toothed comb; and then he was furnished with a suit of new clothes, and taught how to put them on and off. They all emerged from this ordeal as shy as a peacock, just plucked. A new English name finished the preparation for the alphabet and the English language" (Berkhofer, *Salvation and the Savage,* 36; quoting Wilson Hobbs: "The Friends' Establishment in Kansas Territory," Kansas Historical Collections, VIII, 1903–1904, 253). This account makes explicit the connection between the new name and an initiation into English literacy, which is implicit in the Brainerd accounts. The rhetoric of the *Brainerd Journal* lacks the violence of this account, which is not to say that the effect wasn't quite similar. See also Adams 108–112, for controversy over naming at federal boarding schools.

43. In the Bellerophon story (Iliad, Bk. VI), the only mention of writing in Homer, Proetus, suspecting Bellerophon of seducing his wife, sends him on an errand with a letter of introduction, which conveys instructions to kill the bearer. See Roy Harris, *The Origins of Writing* (La Salle, Ill.: Open Court, 1986), 15–16, for one analysis of the story.

44. These details from the *Brainerd Journal*. Worcester is American Board corresponding secretary; his nephew, Samuel Austin Worcester worked with the (Cherokee) Elias Boudinot on the *Cherokee Phoenix* newspaper, earning the Cherokee name A-tse-nu-sti, "the messenger" (*Brainerd Journal*, 396). John Knox (1513–1572) was the founder of Scottish Presbyterianism, while John Witherspoon (1723–1794) was a Presbyterian minister and president of the College of New Jersey; Thomas McKenney was superintendent of Indian trade 1816–1822, and head of the new Bureau of Indian Affairs in the war department from 1824 to 1830; *Boston Recorder* was a newspaper, and the student's benefactor; Lydia Huntley, by 1829 perhaps the most famous poet in America, was a supporter of Indian rights; one of her first books was a narrative poem called *Traits of the Aborigines* excoriating white behavior toward Indians; Jedidiah Morse was a prominent geographer, founder of the Andover Theological Seminary and on the boards of the American Board and the Society for the Propagation of the Gospel; Jeremiah Evarts, corresponding secretary and treasurer of the American Board, was the author of the "William Penn" letters against removal; Elias Boudinot, was a federalist force in revolutionary politics and a benefactor of the Cornwall, Conn., Indian school.

45. *Brainerd Journal* 182 (July 3, 1820).

46. "One of the primary objects of discipline is to fix; it is an anti-nomadic technique" (Foucault, 218). The Cherokee interpreters who translated the names for the mission records had a daunting task, and doubtless the names lost a lot in translation; nonetheless the Cherokee names are suggestive: Taking Away, Man Strikes, Stone Thrower, Bird Pecking, Swimmer, Making Holes Deeper, A Wren Going Under, Run After, Jumping in the Water, Sleeping Rabbit, Let Us Go Across, Standing, Jumper, John Is Gone, Hid, Howling, Hog Shooter, White Killer, Runner, One Who Runs Too Fast to Be Taken (*Brainerd Journal*, 407–418).

47. James Mooney, *The Swimmer Manuscript: Cherokee Sacred Formulas and Medicinal Prescriptions*, ed. Frans M. Olbrechts, Smithsonian Institution Bureau of American Ethnology, Bulletin 99. (Washington: U.S. Government Printing Office, 1932), 127–128.

48. David French and Katherine French, "Personal Names," in *Handbook of North American Indians, Volume 17, Languages*, ed. Ives Goddard (Washington: Smithsonian, 1996), 212.

49. John Howard Payne papers, typescript, vol. 8, 24, 40. Michael Coleman's article on these child-writers as "culture brokers" drew my attention to these riveting letters; "American Indian School Pupils as Cultural Brokers: Cherokee Girls at Brainerd Mission, 1828–1829," in *Between Indian and White Worlds: The Culture Broker*, ed. Margaret Connell Szasz (Norman: University of Oklahoma Press, 1994), 122–135.

50. The project of civilizing had long been associated with western notions of property. Jefferson told the Indians that "when once you have property, you will want laws and magistrates to protect your property and persons. . . . You will find that our laws are good for this purpose"; quoted in Wilcomb E. Washburn, *Redman's Land, White Man's Law a Study of the Past and Present State of the American Indian* (New York: Scribner, 1971), 61. Similarly, William Crawford, secretary of war

under Madison, thought that "The idea of separate property in things personal universally precedes the same idea in relation to lands" (*American State Papers, Class II, Indian Affairs,* Washington, D.C., 1832, 27). They have in mind only particular items of property; specifically, hoes, plows, spinning wheels, looms figure prominently along with slaves, cattle, and pigs in reports of Cherokee progress. See also Ronald Horsman, especially chapter 10, for the relationship between these notions of proper property and the increasing racialization of white and Indian relations; *Race and Manifest Destiny: The Origins of American Racial Anglo-Saxonism* (Cambridge: Harvard University Press, 1981).

Telegraphy's Corporeal Fictions

Katherine Stubbs

The last decade of the twentieth century witnessed the feverish proliferation of scholarship focusing on the internet. Much of this critical excitement sprang from one of the intriguing features of early internet technology: In its first incarnation, online communication was entirely text based. In the eyes of many observers, text-based social interactions seemed to offer a freedom from the constraints of the body; cloaked in the anonymity of print, users could strategically misrepresent themselves to others online. For many, the internet stood as a potentially emancipatory arena, a space in which it was possible to experiment with selfhood; in the words of Sherry Turkle, sites such as internet chat rooms served as "laboratories for the construction of identity."[1] Despite evidence that online masquerades frequently failed, and despite the fact that this type of experimentation often left intact a variety of forms of prejudice, the internet's promise to release users from their corporeal restraints continued to be persuasive.[2] By the end of the decade, the notion that the internet enabled an escape from the limits of the body had become a widespread trope in American culture, present not only in scholarly treatises but in the rhetoric of commercial purveyors of the technology. Thus, in 1997, an MCI television ad could promote the internet by proclaiming that, "People . . . communicate mind to mind. There is no race. There are no genders. There is no age. There are no infirmities. There are only minds. . . . Utopia? No, the internet. Where minds, doors, and bodies open up."

What all of these representations have in common is a belief in the ability of the internet to reconfigure the real according to imagined constructs—to translate the actual into the "virtual." But if the liberating and politically progressive possibilities of this feature of the technology have been well explored, the concept of virtual reality, resting as it does on a problematic dematerialization of the real world, demands further attention. Insofar as we cast the virtual as an abstract realm distinct from material contingencies and conditions, we deny the very realities that enable the technology and in which the technology crucially participates. Strikingly, this discursive strain—the virtual as radically divorced from the real—continues to circulate at a time when the internet has become not only a vast consumer medium but also "the central production and control apparatus of an increasingly supranational market system."[3] While the economic function of the internet has become highly visible, many theorists remain reluctant to assess the ways in which economic considerations have crucially shaped our fantasies, the cultural meanings we have assigned to the technology.

In order to gain perspective on what lies beneath contemporary dreams about media technology, we might find instructive the case of an earlier communication technology, the telegraph. Although at first glance it does not appear to share many features with the internet, the telegraph did in fact raise remarkably similar issues regarding the status of the body and personal identity in relation to technology. A telegraph operator was a member of a community; as many as ten or twelve operators might work on the same telegraph circuit, rapidly transmitting and receiving messages using Morse code. The wire was akin to a party line, as every message transmitted over the wire could be read by all the operators. On certain less-trafficked rural lines, in the intervals when no official telegraph messages were being sent, operators would routinely have personal conversations with each other over the wire. Given the nature of the technology, it was impossible to know for certain from which station a given message originated. The operators on the line were supposed to identify themselves at the beginning of each message, but there was no way to verify definitively the identity of a given sender.[4] The result was a form of anonymity analogous to that enabled by the internet: On the telegraph circuit, it was theoretically possible to misrepresent oneself, to engage in a covert form of masquerade, trying on a new body and a new social identity.

The intriguing parallels between the telegraph and the internet may tempt us to collapse the differences between the two media. But in casting the telegraph as "the Victorian internet," as the critic Tom Standage has done, we risk reducing the telegraph to the status of mere precursor to the later technology.[5] Rather than positioning the telegraph as the lost origin of the internet, we might instead seek to explore the technology on its own terms, investigating how those actually involved in operating the technology conceived of its pleasures and dangers. A little-known genre of literature, the "telegraphic fiction" written by telegraph operators in the 1870s and 1880s, for example, makes the crucial distinctions between the telegraph and the internet, between that historical context and our own, clear. Yet telegraphic fiction also has much to teach us about the curious fantasies in circulation around the communication technologies of the twenty-first century. By exploring these early fictions, many of which seek to investigate the status of the body and personal identity in relation to the telegraph, we begin to glimpse the degree to which representations of communication technologies frequently serve as attempted symbolic resolutions of larger contradictions within American culture.

By the late nineteenth century, when most of this telegraphic fiction was produced, the telegraph was no longer brand new; it had been introduced thirty years earlier. But when we follow Carolyn Marvin's injunction to shift our focus of interpretation from technologies themselves to the social drama in which the technologies are used to "negotiate power, authority, representation, and knowledge," we discover that the telegraphic fictions of the late nineteenth century do in fact seem to constitute a historically new conception of the technology: In a sense, these cultural responses conceive of the technology in new, unprecedented ways.[6] When we read this fiction today, we cannot help but be struck by the stories that focus on the technology as a tool of masquerade. From our contemporary vantage point, fictions such as Edward O. Chase's 1877 tale, "Wives for Two; or, Joe's Little Joke," seem to be eerily prescient about the dangers of using communication technologies to tinker with identity. Set in a New England telegraph office, where every evening there is banter, or "buzzing," between operators over the wire, the story seems nothing less than an avatar of on-line sexual cruising. Chase writes: "Two of the girls would start it, perhaps, with talk of dress and furbelows, shy hints at village beaux and gentle rivalry, then Joe, who was always on the alert for such chances, would snatch the circuit from one of the fair ones, and send a different and perhaps startling answer to the last question."[7] A practical joker, Joe frequently breaks in on discussions, impersonating one of the operators, to create embarrassment. The story's plot becomes tangled when both Joe and the male narrator fall in love over the wire with a female operator named Dolly Vaughan. Joe convinces another female operator, Mabel Warren, to impersonate Dolly, so that the narrator might unknowingly switch allegiance to the second woman. The narrator then visits his new love in person, spending the day with "Dolly" and becoming engaged to her. At the wedding, the narrator discovers "the cream of the joke": When Joe unexpectedly marries the other female operator, the narrator realizes that Joe has wed the real Dolly and he himself has wed an impostor.

On the surface, "Wives for Two" appears to be a slight, if amusing, drama of mistaken identity, a carefully calculated romantic comedy of errors. With the final twist revealed, Chase's tale seems to deliver just what the title advertises—a little joke. Any satisfaction for contemporary readers derives in part from an appreciation of this narrative as a fictional foreshadowing of the forms of identity manipulation promised by later communication technologies. But I would suggest that this story is more than a humorous set piece;

its depiction of an on-line masquerade cannot be considered apart from its specific historical context. Indeed, there is a great deal at stake in the scenario traced by this narrative, although the stakes become evident only when we read the story as a pointed response to a particularly fraught moment in the history of telegraphy.

The history of telegraphy in America is, in large part, the history of an industry. From its introduction in the 1840s, the telegraph was a highly commercial technology; although many other countries sought governmental control over the wires, in the United States the technology belonged to a number of profit-oriented companies. By 1873, however, any competition between them was over, as Western Union, founded in 1856, finally secured a monopoly.[8] With contracts and rights-of-way for most of the nation's wires, the company was conducting 80 percent of telegraph traffic in 1880. In fact, as Edwin Gabler observes, "the telegraph business was synonymous with Western Union" throughout the late nineteenth century.[9]

But while Western Union dominated the market, it was unable fully to control its workers; the company's labor relations were notoriously strained. Western Union had experienced a widespread operator strike in 1870, and it suffered another damaging walkout in 1883. These actions only reminded Western Union management that their industry relied to an uncomfortable degree on its human telegraph operators. Each telegraph transmission involved at least two operators; in order for a message to be transmitted, an operator had first to read and understand it, translate it into Morse code and transmit the dots and dashes over the wire (using the telegraph "key"), where another operator would listen to the coded sequence (coming in through the "sounder") and convert it into comprehensible English. Far from being a purely mechanical relay, transmission depended on the operator, who therefore occupied a highly important position. Writing in 1876, the president of Western Union admitted that telegraphy was distinguished from other means of communication in that it required a human medium: "Presumably no one knows the contents of a . . . letter, except the sender, whereas, anyone sending a telegram of necessity communicates it to another person, the operator. . . ."[10] "The Telegrapher's Song," a poem that appeared about the same time, drove home this point, although in more celebratory language; in that piece, the operator "weave[s] a girdle round the globe / And guide[s] the lightning's wing"; "We touch our key, and, quick as thought, / The message onward flies—/ For every point within the world / Right at our elbow lies."[11]

Because the operator played a critical role in the system of telegraphic communication, labor gained a formidable influence over the industry's monopoly capital and the nation itself. A clandestine organization of operators, the Telegraphers' Protective League, declared in 1868:

> telegraphy's very peculiarities enhance [operators'] facilities for self protection [against the company], and while in its nature it must ever be controlled by a vast capital, its foundation rests in our hands. Each individual operator is a component part of the great system, without which the commercial interests of the country would be paralyzed, were our services withheld for a single week.[12]

Western Union's management was uncomfortably aware of the operator's leverage. Denouncing the operators' strike of 1870, a company spokesman declared, "Practically the intercourse of the continent by telegraph, on which innumerable interests had already learned to depend, was put in peril. . . ."[13] And the company's in-house journal lamented, "the power of operators to annoy and destroy is vast and fearful."[14]

From its very beginnings, industry insiders had attempted to eliminate the operator from the system. Samuel Morse, inventor of the code used throughout the industry, originally attempted to institute a system that would automatically transmit messages; he assumed that such a mechanism was necessary to guarantee the accuracy of the telegraphic transmission.[15] And throughout the late nineteenth century, industry commentators repeatedly called for mechanization of the technology.[16] Yet despite the efforts of the large number of inventors who attempted to introduce automatic telegraph machines, these devices did not go into widespread use until the First World War; in the nineteenth century, most automatic machines proved to be far more delicate than the simple, sturdy Morse model, and very difficult to repair.

If, during the late nineteenth century, it was not feasible fully to automate the system, it seemed crucial to control the operator. For capital truly to control the technology, the laborer had to be reduced to a component of the machine. As a Western Union spokesman argued in the early 1870s, "The telegraph service demands a rigorous discipline to which its earlier administration was unused. The character of the business has wholly changed. It cannot now subserve public interests or its own healthful development without the precision and uniformity of mechanism."[17] What was needed was a mediating agent who was not an agent at all; the ideal operator would be inexpensive and easily subjected to management's regime of "rigorous discipline." This desire to control more closely the unpredictable human operator helps explain how the telegraph industry came to prefer women as operators. Female operators were assumed to be less likely to agitate for better working conditions and higher wages, less likely to strike, and more amenable to management's disciplinary strategies. Indeed, management tended to cast the female operator as an inert element of the communication apparatus. Thus it was not simply that female operators could be hired at lower wages; they were also valued for their putative docility, their perceived willingness to be servants of the technology and the company.[18]

By the 1860s, Western Union's preference for female operators was unmistakable. In 1869, the company began a joint venture with the Cooper Union Institute in New York, running an eight-month telegraphy course; the explicit goal of the school was to train female operators. The institute's first annual report phrased it bluntly: "The experience of the telegraph companies has gradually but surely convinced the managers that their interests would be greatly promoted by the substitution of women for men in the greater number of offices."[19] A member of the Buffalo, New York, Western Union force, writing in the early 1880s, noted that it was "understood that the policy of the manager now is to fill all vacancies with ladies—at about one-half the price formerly paid, of course."[20]

Yet if management widely predicted that "ultimately a large proportion of the telegraphists . . . would be females," as a superintendent declared in 1863, the process of feminization was very gradual and did not go uncontested.[21] As that manager acknowledged, the main obstacle to feminization was "the antagonism naturally felt by male operators, who see in it a loss of employment to themselves. . . ."[22] Indeed, male operators were acutely aware that the number of female operators was steadily increasing. One operator pessimistically predicted that by 1923, male operators would be completely eliminated: "The American Rapid and multiplex/Will be worked alone, by the gentle sex;/The men who now at the key do toil/Will have to die or till the soil."[23]

Telegraph operator culture in the late nineteenth century can thus be seen in part as a response to this perceived crisis of feminization. Male operators reacted to this crisis in a variety of ways. Some addressed the issue directly, by writing editorials in industry papers or by making scornful pronouncements such as Jasper Ewing Brady's confession, "I wasn't overfond of women operators."[24] But a far more revealing response surfaced in the 1870s and 1880s, in the stories and poems known as telegraphic literature. This literature appeared primarily in industry newspapers and was produced and consumed by telegraph operators, although it was also available to the public in the form of published story collections.[25] One of the chief venues for telegraphic literature, the industry paper the *Operator*, published from 1874 to 1885, provided a forum for the laborers to express themselves. The paper conceived of itself as the only "thoroughly independent" trade journal in the industry; it advertised that it was "devoted to the interests of the Telegraphic Fraternity," and "not controlled by any corporation or company."[26] As the emphasis on "fraternity" suggests, contributors to the *Operator* were overwhelmingly male, and virtually all were operators themselves.[27] The journal's poems and stories, together with the tales written by the operator Walter P. Phillips, can be interpreted as a clear attempt to combat Western Union's efforts to feminize the telegraph industry.

Many of these writers adopted the basic tactic of romanticizing the male operator. In the very early years of the industry (before and during the Civil War), male operators who entered the field had often ascended through the ranks to managerial positions. But in the last three decades of the nineteenth century, as the industry became more rigidly organized, opportunities for advancement within Western Union diminished sharply.[28] As Gabler notes, "low incomes and crippled mobility" and "narrowed skill ranges and status" combined to make those years "the dark age of the craft" for operators.[29] Many telegraph operators in the late nineteenth century came to view their work in post-lapserian terms, saturating it with a fierce nostalgia for what was considered the industry's golden age, the years before the 1870s. According to this view, telegraphy was once a true craft, practiced by mythic men known within the field as "boomers," who were famous both for their skillful work and for their irresponsible debauchery. Writing under the nom de plume John Oakum, Walter P. Phillips commemorated the boomers' charming eccentricities in a series of stories published in the industry papers.[30] But if the image of the boomer pleased some operators, it worried others. At a time of industry rationalization, some male operators felt the need to dispute publicly the stereotype of the footloose, fast-living boomer, by replacing this rogue figure with the image of the operator as an upstanding hero and moral exemplar—a chivalrous "knight of the key," in the common phrase used by male telegraphers.[31] Although the "knight of the key" stereotype differed from the boomer image, both myths nonetheless presented the operator as a virile male, and envisaged the telegraph industry as a highly masculine, homosocial realm.

While glorifications of the fraternity of operators worked implicitly to exclude female operators, there were also numerous representations that addressed the threat of female operators more explicitly. The writers of several works of fiction went to great lengths to stress that the introduction of female operators would result in a decline in operator competence. In such tales, female operators are depicted as woefully lacking in the skills necessary to operate the machines; often the work of male operators also suffers because they are distracted by women working nearby.[32]

But a far more intriguing and complex literary response to feminization surfaced in another set of tales, stories that can be read as veritable meditations on the relation between the human body and communication technology. In these tales, the proximity of the female body to the telegraph machine alters the technology itself, effectively disabling it. J. M. Maclachlan's story "A Perilous Christmas Courtship; or, Dangerous Telegraphy" invokes just such a scenario. When the main character of Maclachlan's tale falls in love with a female operator who is willing to "'do a flirt' on the wire," the result is a highly eroticized encounter.[33] The main character declares, "Such a burning stream of affection, solicitude, and sentiment flowed over that senseless iron thread, when not ruthlessly interrupted by common-place dispatches, that I often thought, when our words grew warmer than usual, that the wire might positively *melt,* and so cut the only link that bound us in love together!"[34]

Despite the hyperbole of Maclachlan's vision, his tale serves as a variation on what was a widespread trope of telegraphic fiction: a depiction of the female operator as a threat to the technology. In some stories, such as William Lynd's "A Brave Telegraphist" and J. A. Clippinger's "Poor Dick," the presence of the female operator disables a crucial function of the technology: the confidential transmission of information. Other tales, including J. M. Maclachlan's "A Perilous Christmas Courtship," Exie's "Foiled by a Woman's Wit," and William Lynd's "A Brave Telegraphist," depict the female operator in dire peril, subject to a variety of brutal assaults. By repeatedly emphasizing the female operator's vulnerability to violent penetration, these stories suggest that the female operator is a weak link in the telegraph network, a theme that is crucial to male operators' arguments against feminization. (The only logical way to restore integrity to the telegraph network, the argument goes, was to remove the female operator from the industry; in numerous stories, this is accomplished through marriage.)[35]

In a large number of these representations, the female operator's weakness—her vulnerability, her assailability—is a function of the degree to which she is exposed to the public. Throughout the nineteenth century, wage-earning women of any sort were considered morally suspect precisely because they promiscuously circulated in unregulated public spaces, outside the safety of their putative natural place in the private sphere of hearth and home.[36] At a time when even the most minimal contact with the public sphere seemed to threaten feminine modesty, female telegraph operators appeared to herald a new and unprecedented form of public exposure. Indeed, as the historian Sarah Eisenstein has noted, telegraphy was one of the most highly public occupations open to women during this period.[37] Male-authored representations of female operators thus often traded on nineteenth-century beliefs about public exposure and female sexuality. In Thomas C. Noble Jr.'s story "Old Robin York," in William Johnston's "A Centennial-

Telegraphic Romance," and in G. W. Russell's poem "Out of Adjustment," the female operator emerges as morally loose—flirtatious at best, sexually indiscriminate at worst. The young female operator Tilly M'wens ruefully sums up these sentiments in the story "A Quandary": "Telegraphing's too public."[38]

Seen in the light of these representations of the female operator, and in the context of the contested gender relations that divided the telegraph labor force during the 1870s, the identity masquerade limned in Edward O. Chase's "Wives for Two; or, Joe's Little Joke" begins to look rather different. The facetiousness of Chase's sketch gives way to a serious subtext. Indeed, Chase's entire narrative is structured by its final twist, the punchline of the joke—the spectacle of the male narrator successfully hoodwinked by a "subtle exchange." This carnivalesque upheaval of expectations, a destabilization of the order to which the narrator was accustomed, is a source of humor, but also of anxiety. "Wives for Two" in effect emphasizes the ways in which the telegraph might be turned into a duplicitous device directed against the male operator. Throughout the tale, Chase relies on the reader's acquaintance with stereotypes about morally loose female operators—for if Joe is the mastermind, Mabel Warren, who serves as his willing accomplice, successfully executes his deception.

To recognize that "Wives for Two" is a cautionary tale, and not, as it would first seem, a simple validation of the transformative possibilities offered by the telegraph, is not to deny that there existed early celebrations of this aspect of communication technology. But it does mean that we must carefully consider how fantasies about the technology reflect historically specific conditions, especially the economic and social tensions of the moment. Ella Cheever Thayer's 1880 novel *Wired Love: A Romance of Dots and Dashes* offers a case in point. Perhaps the most extensive nineteenth-century meditation on the ways communication technology can be used to negotiate embodiment, *Wired Love* was published by W. J. Johnston (the editor of the *Operator*), and was heavily promoted in that journal as the "first and only telegraphic novel."[39] It depicts the adventures of Nattie Rogers, a young female operator who falls in love over the wire with an operator known only as "C" (later revealed to be a young man named Clem). Over the course of the novel, Nattie suffers several setbacks in her pursuit of Clem (as "Cupid, viewed in the character of a telegraphist . . . switche[s] everybody off on to the wrong wire") before ultimately achieving her goal, when the "circuit" of her love for Clem is completed.[40] But despite its conventional structure, this romantic narrative is far from traditional. In fact, Thayer's novel portrays the telegraph circuit as a space that enables corporeal and social plasticity.

Throughout her novel, Thayer responds to those who would indict the female operator for her publicity. Rather than denying the publicity of the female operator's body, she instead begins by strategically foregrounding it. During much of the novel, Thayer uses Miss Kling, the landlady of Nattie's boardinghouse, as a spokeswoman for conventional expectations of female modesty (she offers statements such as "I was brought up to understand that young ladies should never receive the visits of gentlemen except in the presence of older people!" [168]). The landlady also repeatedly articulates the standard condemnations of female publicity. "Public characters are not to be trusted," she argues (64). "You cannot deny that no young woman of a modest and retiring position would seek to place herself in a public position," she declares (65). Another character later echoes Miss Kling's complaints, crying, "How can she appear before the public so? [I]t seems so unwomanly!" (186). While the explicit target of these comments is a young woman named Cyn, an opera singer, Thayer makes clear that Nattie is equally vulnerable to charges of publicity. Indeed, throughout the text, whenever Nattie is in the telegraph office, Thayer stresses her visibility.[41]

As if the reader might miss the moral implications of Nattie's publicity, Thayer takes pains to accentuate her heroine's aggressive pursuit of Clem. Although Nattie sees Clem every day in the boardinghouse where they both live, she nonetheless becomes jealous of her friend Cyn's close relationship with Clem. Nattie is entranced with a new idea: the construction of a private telegraph wire from her bedroom to the bedroom where Clem lives with his roommate. That way, she will get Clem all to herself and "*nobody* could share" him (175). Thereafter, Nattie and Clem achieve an intimacy highly unconventional for unmarried young persons in the nineteenth century: Each morning, Nattie performs the duty of a wife to Clem, waking him up by calling him over the private line; each night, Nattie and Clem use the line for pillow talk, chatting "even into the small hours" (180).

Inevitably, Miss Kling suspects "a horrible scandal" to be taking place, and finally confronts Nattie with the "underhanded" "utter immorality" of "such proceedings" (245). In the same sentence that she condemns women who "show themselves to the public" she denounces Nattie's private line to Clem's bedroom (245). She declares, "That any young woman should be so immodest as to establish telegraphic communication between her bed-room and the bed-room of two young men is beyond my comprehension!" (246). In response to Miss Kling's accusation that she has "connected" herself to Clem, Nattie blushes and confesses that while the private line was "not exactly correct," "there was nothing reprehensible in my conduct" (246). The landlady then responds in the strongest terms. "I have heard of young females so much in love that they would run after and pursue young men, but never before of one so carried away and so lost to every sense of decorum, as to be obliged to have a wire run from her room to his, in order to communicate with him at improper times!" (248). But Thayer immediately makes clear that Nattie will not be punished for her transgression; instead, she is rewarded, as Clem proposes marriage on the spot. Miss Kling comments acerbically to Nattie's lover, "Well, she has worked hard enough to get you—had to bring the telegraph to her assistance!" (252). As the novel's happy ending demonstrates, Thayer decisively condones Nattie's violation of the normal rules of courtship.

Thayer is able to offer this approbation because she constructs the telegraph circuit as a space different from the space of daily life; Nattie "lived, as it were, in two worlds" (25). Her "telegraphic world" (26) enables her to "wander away, through the medium of that slender telegraph wire, on a sort of electric wings, to distant cities and towns" (25). This telegraphic world is one "where she could amuse herself if she chose, by listening to and speculating upon the many messages of joy or of sorrow, of business and of pleasure, constantly going over the wire" (25). But Nattie is particularly thrilled by the "romantic" potential of the telegraphic circuit, because while she is on the circuit, previously unsanctioned forms of social connection can take place; Nattie's private telegraph wire is only one of many instances of this. When Nattie first becomes acquainted with "C" over the wire, she permits him liberties that would be unacceptable in real life. She quickly becomes accustomed to "the license that distance gave," so that when "C" addresses her impertinently as "my dear," Nattie excuses him by thinking, "did not the distance in any case annul the familiarity?" (41). And when "C" expresses his hope that "we may clasp hands bodily as we do now spiritually, on the wire" (44), she readily assents.

Indeed, Nattie's most intimate moments with "C" always occur through the medium of telegraphy. But what is most astonishing is that Thayer presents Nattie and Clem as entirely dependent upon the circuit for their interaction, even when they are face to face.[42] At such moments, in order to express themselves freely, they pretend to be connected through a phantom circuit, and use objects to drum out a message in Morse code. In their first face-to-face meeting, "C" reveals himself to Nattie in code by using a pencil to tap out a message. Nattie responds in code, using a pair of scissors (148–151). Whenever a spare telegraph machine is available, they stand facing each other, tapping out messages. Clem acknowledges, "It *is* nicer talking on the wire, isn't it?" (173), and declares, "a wire is so necessary to our happiness!" (174). As Cyn notes, even if Clem and Nattie were stranded by themselves on a desert island, a telegraph wire would be needed for their communication. It is fitting, then, that *Wired Love* concludes with Nattie and Clem standing in the same room, hands clasped over the same telegraph key, telegraphing their love to each other (256).

The telegraph circuit is the sole arena where Nattie can express herself freely because to speak on the circuit is to create a new identity which is not identical to one's identity in the real world. Thus, after getting to know Clem, Nattie feels a sense of loss and nostalgia for "C": "She sometimes felt that a certain something that had been on the wire was lacking now; that Clem, while realizing all her old expectations of 'C,' was not exactly what 'C' had been to her" (170). Nattie notes that "the day that had brought Clem" in person to her, had "not restored as she then supposed, but taken away, her 'C'" (223). Nattie's acknowledgment of the disparity between "C" as she imagined him and the flesh-and-blood Clem is another way of noting that telegraphy disconnects the speaking self from the body. In the words of Quimby, another of Thayer's characters, telegraphy is a domain where persons become "invisibles" (53, 55, 82). Quimby illustrates his point "with a gesture of his arm that produced an impression as if that member had leaped out of its socket" (82). With this eloquent representation, Quimby locates his commentary at the level of the body; if the arm leaving its socket figures a corporeal alienation (a body divided against itself), it also suggests an attempt to escape the body. Or, in another of Quimby's formulations, telegraphy is "ghostly" (82) not only because it was conventionally assumed to be a link to the spiritual world, but also because it enables an escape from the body, a form of self-abstraction.

This insight functions as the central contribution of *Wired Love*. Traditional republican rules of public discourse maintained that the public subject's right to speak was predicated on his disembodiment and anonymity; he could assume the speaking position of the disinterested public citizen only because he inhabited a body perceived as so normative (white, male, upper-class) that it did not count as a body at all, a concept that Michael Warner has termed "the rhetorical strategy of personal abstraction."[43] As Warner argues, this ability to abstract oneself, to take rhetorical advantage of the nonidentity of speaking self and embodied self, has not historically been equally available to everyone. For those persons whose bodies deviated from the white male standard, corporeal difference was equivalent to hyper-embodiment; unable to abstract themselves, they had no grounds from which to speak in the public sphere. To attempt to speak would only call attention to their bodies, to their lack of authority; for women, this meant that appearing in too public a fashion was a form of dangerous promiscuity. Thus, in the male-authored fantasies of telegraphic literature, the female operator, exposed to multiple forms of public circulation, is not disembodied and anonymous, but is instead all too visible, precariously hyper-embodied.

If, as Warner has argued, in the eighteenth century the public sphere was constructed through print discourse, we see in Thayer's novel the possibility that a new form of public sphere is created by the circuit of communication technology. In Thayer's novel, the telegraph enables self-abstraction, a nonidentity with the body. This disincorporation is not a denial of the female operator's publicity, but a reclamation of it from those versions that would equate the female body with the dominated body. In the space of the circuit, then, the female operator experiences a different relation with herself; she earns the right to speak.[44] Rendered invisible, temporarily freed from her body, she is also freed from conventional rules of female behavior and seems no longer subject to the traditional forms of discipline, prejudice, and violence that exploit corporeal difference.

The writings of two female operators, J. J. Schofield and L. A. Churchill, extend Thayer's notions regarding the benefits communication technology offered to nineteenth-century women. Along with Thayer, Schofield and Churchill joined the small handful of female writers of telegraphic fiction; Churchill was the only female contributor to the first edition of the fiction collection *Lightning Flashes* and was joined by Schofield in the second edition of the text (where, in the table of contents, the two writers were distinguished from their colleagues by the appellation "Miss"). In stories by Schofield and Churchill, the telegraph circuit appears to emancipate the female operator from visibility and corporeal entrapment, from aspects of her body that marginalize her. Specifically, the circuit becomes a way of circumventing cultural expectations regarding age, physical appearance, and female sexuality.

In Schofield's "Wooing by Wire," Mildred Sunnidale, a telegraph operator, believes herself handicapped in romantic encounters because of her physical appearance and her advanced age (unmarried at thirty, she qualifies as a "spinster," as she acknowledges). Mildred establishes a relationship over the wire with a male operator, Tom Gordon. Tom is not only kind—he forces abusive male operators to treat Mildred respectfully—but he is also enlightened; his experiences over the circuit have taught him that physical appearance is unimportant. As he tells a friend, "You can form quite as good an estimate of a girl's character and temper by working and talking with her over the line, as by being personally acquainted with her; better, perhaps, for you take her on her merits alone, and are not prejudiced by appearance."[45] Tom has thus learned, after his experiences on the telegraph circuit, that it is possible to see beyond the exterior appearance of "homely women," to stop "taking it for granted that their characters are as unattractive as their faces" (164). According to Schofield's representation, Mildred's relationship with Tom is possible largely because Tom cannot see Mildred's face, and because he fails to realize how old she is (she speculates, "Would he want [my photograph] if he knew I was thirty years old, and so dreadfully plain?" [167]). When Tom proposes to Mildred over the wire, she tells him, "When you saw me, and discovered your mistake, you would repent your bargain. Why, I'm an old maid, with red hair!" (167). But Tom has already fallen in love with Mildred and thus still wants to marry her even after he has seen her. As he acknowledges at the conclusion of the story, he would never have discovered her good nature had he first met her in person; his love for her would not have had a chance to develop "except by working with her over the line" (168).

The terms of Schofield's story are echoed in L. A. Churchill's "A Slight Mistake." In Churchill's tale, a young operator named Paul Riverson has fallen in love with an operator named Flossie Bates. Flossie, "a woman of about fifty, with a decidedly stout figure and a profusion of gray hair," is thrilled with her new relationship, with the way the circuit frees her from conventional prejudice about age and appearance.[46] As long as the two communicate over the wire, the romance flourishes. But Churchill does not have Schofield's optimism; at the conclusion of the story, when Paul sees Flossie in person, he is outraged at his "mistake." He immediately rejects Flossie and becomes embittered, vowing never again to risk flirting over the wire.

"Wooing by Wire" and "A Slight Mistake" both explore the use of the circuit to make one's body metaphorically pliable, to attempt to establish a new social identity and new social relations based on that identity. Perhaps the most striking instance of such a masquerade appears in L. A. Churchill's "Playing with Fire." The story depicts a female operator named Rena Chelsey, who resolves to fool another female operator at the next telegraph station, a Miss Dwinell, by pretending to be a man, "Isaac." In that persona, she flirts with Miss Dwinell, complimenting her on her appearance (she has in fact glimpsed Miss Dwinell one day from afar). After a day of extensive flirtation, Rena declares to herself, "One would suppose from my talk that I was a regular lady-killer."[47] Over the following weeks, Rena and Miss Dwinell spend many hours together on the wire. It is soon apparent that Miss Dwinell has fallen in love with Rena, and Rena too is deeply stirred by Miss Dwinell's "charm"; she sends her numerous small gifts. When Miss Dwinell announces that she will visit Rena in person, Rena panics. In a telling passage, she both repents of the indecency of her ruse and wishes that she might become a man, so that she could consummate her relationship with Miss Dwinell. "If I had ever imagined that she was half as interesting as I have found her, I would have tried to make her acquaintance at once in a decent way. I shall be heartily sorry to lose her good opinion, for I am so deeply interested in her that I more than half wish this Isaac business was a reality" (70).

The strong homoerotic charge of the story appears defused when Miss Dwinell visits Rena in person and it is revealed that the operator Rena assumed to be Miss Dwinell is in fact a young man. Herbert Stanley, replacing Miss Dwinell in her office, "in the spirit of mischief," decided to impersonate Miss Dwinell when "Isaac" first began flirting. But it immediately becomes evident that this development does not reduce but rather intensifies the queerness of the situation. For Herbert confesses that he has fallen in love with "Isaac." He tells Rena, "You have yet to learn how much I have come to care for my new friend. Often I have wished I *was* Miss Dwinell, if Isaac would care for me as he seemed to care for her" (71). After admitting his upset at Rena's gender ("There must be some mistake" [70]), he makes a half-hearted attempt to assert that "things are now just as they should be." He declares, "I think you cared for me in my old character. Can you learn to love me in my new?" But Rena, it turns out, is also disappointed, and refuses his offer. "Home love is not for me," she informs Herbert, and advises, "Let us each go our own way, bravely walking in the path marked out for us . . ." (71). In the last lines of the story, Churchill explains that neither Rena nor Herbert ever marry. Herbert feels that he has another unnamed "purpose to fulfill" (71), as does Rena. Churchill concludes the tale by reflecting not on the pathos of Rena's spinsterhood but on the dignity of her destiny: "A few years later Rena died. Her work was done. The brave heart ceased to beat, and the tired hands were folded, and no one said that hers had been a wasted life" (71). With this, Churchill suggests that there can be no tidy heterosexual solution; the circuit has permanently queered things for Rena and Herbert.

If the female-authored fictions of Churchill, Schofield, and Thayer revel in technology's offer of emancipatory subjectivity to those who were highly circumscribed in the nineteenth century, it is worth remembering that many male-authored fictions were preoccupied with the threatening side of this aspect of the technology, its potential to endanger the status of those who were dominant. Recognizing the range of responses to this aspect of the technology—that there existed representations of masquerade that were censorious, as well as celebratory—prevents us from positing a straightforward genealogy of communication technologies, whereby fantasies about the internet simply mirror earlier fantasies about the telegraph. It is wise for us to recognize that these early fantasies were responses to a specific historical moment; the tales of telegraphic fiction offered symbolic resolutions to very immediate and urgent tensions within nineteenth-century society. One way to read these tales is as efforts to reconcile traditional assumptions about gender roles—about men and women's proper and rightful place—with new workplace conditions, with rapidly changing economic and social realities.

Yet if in one sense fantasies about telegraphy speak narrowly to their own historical moment, they are also suggestive of ours. In telegraphic fiction, what is all too often mystified in current discourse about the internet is rendered relatively accessible. In these nineteenth-century fictions, the material conditions underlying the operation of technology become visible. While the telegraph is often fantasized as a means of achieving temporary release from certain constraints, it just as often appears an as integral participant in the reproduction of unequal conditions and social relations. With the example of telegraphic literature before us, we might turn to our contemporary moment, to interrogate the cultural meanings we have assigned to the internet, and to inquire about what needs these meanings have been made to serve. It is only then—when we fully acknowledge the degree to which the computer internet is enabled by and itself enables structures of economic exchange—that we begin to recognize that our most persistent fictions about the internet, our fantasies about escape and our longings for fluidity of identity, might also be understood as powerful evidence of the forms of circumscription and impoverishment experienced by those living under the conditions of late capitalism.

Notes

1. Sherry Turkle, *Life on the Screen: Identity in the Age of the Internet* (New York: Simon and Schuster 1995), 184.

2. It is important to note that the invisibility of corporeal difference does not always lead to politically progressive results; insisting on the visibility of alterity in cyberspace can serve as a mechanism through which to interrogate the reproduction of normative assumptions in real life. Indeed, a denial of one's difference can be problematic in a variety of situations, including those involving community building. Julian Stallabrass has argued that "the extreme mutability and multiplication of identity possible in cyberspace collides with the desire to build communities based upon honest communication with people of diverse backgrounds and interests. Role-playing, and the potential for dishonesty which goes with it, militates against community." "Empowering Technology: The Exploration of Cyberspace," *New Left Review* 211 (May/June 1995): 16. See also Christina Elizabeth Sharpe, "Racialized Fantasies on the Internet," *Signs* 24, no. 4 (1999): 1089–1096, and Jodi Dean, "Virtual Fears," *Signs* 24, no. 4 (1999): 1069–1078.

3. Dan Schiller, *Digital Capitalism: Networking the Global Market System* (Cambridge: MIT Press, 1999), xiv.

4. Although some commentators and operators believed that certain operators—particularly women—possessed a characteristic sending style, such attempts at distinguishing the true identity of a given sender were notoriously difficult.

5. Tom Standage, *The Victorian Internet: The Remarkable Story of the Telegraph and the Nineteenth Century's On-line Pioneers* (New York: Berkley Publishing Group, 1998).

6. Carolyn Marvin, *When Old Technologies Were New: Thinking About Electric Communication in the Late Nineteenth Century* (New York: Oxford University Press, 1988), 5.

7. Edward O. Chase, "Wives for Two; or, Joe's Little Joke," *Lightning Flashes and Electric Dashes: A Volume of Choice Telegraphic Literature, Humor, Fun, Wit & Wisdom* (New York, 1877), 41.

8. Vidkunn Ulriksson, *The Telegraphers: Their Craft and Their Unions* (Washington, D.C.: Public Affairs, 1953), 8. Western Union's only competition came from the Baltimore and Ohio Telegraph Company and the American Rapid Telegraph Company.

9. Edwin Gabler, *The American Telegrapher: A Social History, 1860–1900* (New Brunswick: Rutgers University Press, 1988), 44, 43.

10. "Inviolability of Private Dispatches," *Journal of the Telegraph,* 1 July 1876, 201.

11. J. A. Wyllie, "The Telegrapher's Song," *Lightning Flashes and Electric Dashes: A Volume of Choice Telegraphic Literature, Humor, Fun, Wit & Wisdom* (New York, 1877), 75.

12. Quoted in Gabler, *The American Telegrapher,* 148.

13. James D. Reid, *The Telegraph in America: Its Founders, Promoters, and Noted Men* (1878; New York, 1974), 547. In his position as a company official and editor of the company organ the *Journal of the Telegraph,* Reid served as a mouthpiece for company doctrine.

14. Quoted in Gabler, *The American Telegrapher,* 157.

15. Charles L. Buckingham, "The Telegraph of To-Day," *Scribner's Magazine* (July 1889), 4. Morse originally conceived of his code as visual; the received messages would be recorded permanently on paper (enabling an objective way of verifying the accuracy of message transmission). He was angered when he realized that operators invariably elected to listen to the pattern of clicks made by the machine, rather than reading the code off a ribbon.

16. Menahem Blondheim has studied the shift in public attitudes toward the telegraph and notes that while initially observers cast the technology in utopian terms, as the century wore on there was increasing anxiety about the degree to which the technology was vulnerable to manipulation by criminals—and by the telegraph companies themselves. Menahem Blondheim, "When Bad Things Happen to Good Technologies: Three Phases in the Diffusion and Perception of American Telegraphy," in *Technology, Pessimism, and Postmodernism,* ed. Yaron Ezrahi, Everett Mendelsohn, and Howard Segal (Dordrecht: Kluwer 1994).

17. Quoted in Gabler, *The American Telegrapher,* 67.

18. See Charles H. Garland, "Women as Telegraphists," *Economic Journal* 11 (June 1901): 254, 255, 260, 261.

19. Quoted in Gabler, *The American Telegrapher,* 132.

20. Quoted in ibid.

21. The feminization of telegraphy during this era was of course being paralleled by the feminization of clerical work. Gabler notes that male fears of industry feminization "were reasonable

enough, given the state of the craft in the postbellum years," although the total feminization of the field did not occur until the twentieth century. The official policy of the national union, the Brotherhood of Telegraphers, was to welcome women into the union and to demand equal pay for women; but, as historians invariably note, demands for equal pay for women were usually equivalent to demanding the removal of women from the field, for their chief appeal for employers was their cheap wages.

22. Quoted in Virginia Penny, *The Employments of Women: A Cyclopaedia of Woman's Work* (Boston, 1863), 101.

23. Anonymous, "A Modern Mother Shipton," *Operator,* 23 December 1882, 681.

24. Jasper Ewing Brady, *Tales of the Telegraph: The Story of a Telegrapher's Life and Adventures in Railroad Commercial and Military Work* (Chicago, 1900), 185. For an example of an editorial explicitly addressing the issue, see "Women as Telegraph Operators," *Operator,* 15 May 1882, 198. The article provides an overview of arguments against female operators by exploring the European debate on the subject. Despite the author's recognition of the adequacy of female operators in the United States, the rhetorical weight of the piece lies with the lengthy explications of arguments against female operators. Thomas Edison, himself a former telegraph operator, went on record denouncing female operators, alleging that they did not have "commercial instinct and judgment" "and can't acquire them." Quoted in Gabler, *The American Telegrapher,* 136.

25. For examples of collections, see Walter P. Phillips, *Sketches Old and New* (New York, 1897) and Phillips's volume written under the pseudonym John Oakum, *Oakum Pickings: A Collection of Stories, Sketches and Paragraphs* (New York, 1876). Phillips's 1897 volume was based on stories originally written in the 1870s. See also the 1877 anthology *Lightning Flashes and Electric Dashes,* collected from the pages of the *Operator* by the paper's editor, W. J. Johnston, which went through two editions.

26. These phrases of self-promotion appeared in an advertisement at the back of *Lightning Flashes* and in an advertisement in the *Journal of the Telegraph,* 1 May 1877, 139.

27. See for instance the biological profiles on the well-known telegraphic fiction writers Edward O. Chase (a former chief operator), D. C. Shaw (a chief operator in Portland, Maine), and Joseph Christie (for years an operator in the Philadelphia Western Union office) in *Lightning Flashes.* Although the fact that some writers use only first initials would seem to make their gender ambiguous, in a later edition of *Lightning Flashes,* a distinction is made between male and female contributors by appending "Miss" to the names of two contributors, L. A. Churchill and J. J. Schofield. I will discuss the contributions of these female authors, below.

28. Gabler, *The American Telegrapher,* 63.

29. Ibid., 64, 71, 63.

30. See Phillips's depictions of "Old Jim Lawless," "Little Tip McClosky," "Cap De Costa," "Posie Van Dusen," "Old George Wentworth" and "Patsy Flanagan," in Oakum, *Oakum Pickings* and Wal-

ter P. Phillips, *Sketches Old and New.* Phillips gave his tales an elegiac cast; he set his stories "in the far-off realm of those good old times, when it was thought something smart to work telegraph lines." *Oakum Pickings,* 64. As a reviewer of Phillips's tales noted, in an advertisement appearing in the *Journal of the Telegraph,* they dramatized a breed of "characters now almost extinct," 1 October 1876, 299. Writing the next year for the *Journal of the Telegraph,* a writer mourned the boomers in humorous terms. He recalled those operators "whose personal exploits and achievements buried those of Munchausen in a grave of ignominious tameness; men who would have freely divided their last cent with any unfortunate, but who unluckily, for the unfortunate, seldom had a last cent to divide, but the *tout ensemble* of whose characteristics, had they been described by some one eloquent, would have been compressed, perhaps, into two words—Smart Alecks." *Journal of the Telegraph* , 1 May 1877, 129.

31. Perhaps the most extended version of the myth of male operator as moral examplar is offered in Brady, *Tales of the Telegraph.*

32. For literary explorations of female operators' incompetence, see John Oakum (Walter P. Phillips), "An Autumn Episode" and "Narcissa" in *Oakum Pickings.* For examples of the disruptive effects of female operators on male operators, see Thomas C. Noble Jr., "Washington Butterfield's Experience with Female Help," *Operator,* 1 December 1883, and "Washington Butterfield's Experience with Female Help, Conclusion," *Operator,* 15 December 1883.

33. J. M. Maclachlan, "A Perilous Christmas Courtship; or, Dangerous Telegraphy," *Lightning Flashes,* 66.

34. Maclachlan, "A Perilous Christmas Courtship," 66.

35. See for example Anonymous, "A Quandary," *Operator,* 15 November 1883, 190; Maclachlan, "A Perilous Christmas Courtship," Chase, "Wives for Two," and William Lynd, "A Brave Telegraphist," *Operator,* 15 March 1884, 78.

36. See Thomas C. Noble Jr., "Old Robin York," *Operator,* 15 October 1883, 158; William Johnston, "A Centennial-Telegraphic Romance," *Lightning Flashes,* 107–108; and G. W. Russell's poem "Out of Adjustment," *Lightning Flashes,* 62–53; and especially "A Quandary," where the female operator's "too public" status is a source of anxiety to the man courting her.

37. Sarah Eisenstein, *Give Us Bread but Give Us Roses: Working Women's Consciousness in the United States, 1890 to the First World War* (London: Routledge & Kegan Paul, 1983), 83. Western Union was aware of the possibility that female operators' exposure to the public could be perceived as a problem. It was Western Union practice to segregate women in the large branch offices, or to surround them with eight-foot partitions to shield them from male eyes; Gabler, *The American Telegrapher,* 111. James D. Reid, spokesman for Western Union, insisted that far from introducing an element of sexuality into the telegraph office, a female operator "stopped vulgarity"; her influence "was in every way healthful," *The Telegraph in America,* 171. In their seclusion, their removal from the world, the operators could be presented as isolated from the meaning of the messages they transmitted. As one anonymous writer in the Western Union company organ put it, female op-

erators who are removed from the world "might for aught they feel in the matter, be accelerating communications from one star to another." "The Female Telegraphists at the London Office," *Journal of the Telegraph,* 16 August 1877, 242.

38. Anonymous, "A Quandary," 190.

39. The question of whether Thayer was herself an operator is difficult to resolve. The level of detail with which the novel describes not only Morse code but also a wide variety of conventions of telegraph operation (including numerous telegraphic abbreviations, where numbers and letters serve as industry code), suggest that Thayer did work as a telegraph operator. She certainly was a reader of telegraphic fiction; several of the anecdotes about telegraphy related in the novel were plagiarized from W. J. Johnston's "Some Curious Anecdotes of the Wire" (which appeared in the second, expanded edition of *Lightning Flashes*). Reviews of *Wired Love* fail to shed light on this question; for instance, the *Boston Journal*'s observation that Thayer "is evidently familiar with the electric telegraph" indicates that the reviewer was not certain of her professional status. The review appears in an advertisement for the novel, published in the *Operator,* 15 October 1883, 165. It seems fair to conclude that if Thayer was not herself an operator, she was closely associated with the industry.

40. Ella Cheever Thayer, *Wired Love: A Romance of Dots and Dashes* (New York, 1880), 194, 198. All further page numbers will be included parenthetically in the text.

41. For example, an "urchin, flattening his nose against her window-glass" watches Nattie (62–63). When Nattie exclaims loudly, "a passing countryman stopped . . . to stare" (96). When she is sad she presents "no laughing face to the curious passers-by" (94). The clerk in the store opposite Nattie's office constantly monitors Nattie's behavior; he observes her "for some weeks" (48, 76, 77, 94). At one point, the clerk is joined by another man, and together the two curiously stare at Nattie (77).

42. This element of Nattie's relationship resembles the real-life experience of Thomas Edison, when he was courting his wife. In his diary he recalled, "I taught the lady of my heart the Morse code, and when she could both send and receive we got along much better than we could have with spoken words by tapping out our remarks to one another on our hands," quoted in Standage, *The Victorian Internet,* 142.

43. Michael Warner, "The Mass Public and the Mass Subject," in *Habermas and the Public Sphere,* ed. Craig Calhoun (Cambridge: MIT Press, 1992), 382.

44. We might thus see these fictions as confirming Carolyn Marvin's observation that the history of electric media "is less the evolution of technical efficiencies in communication than a series of arenas for negotiating issues crucial to the conduct of social life; among them, who is inside and outside, who may speak, who may not, and who has authority and may be believed"; Marvin, 4.

45. This story appeared only in the second, expanded edition of *Lightning Flashes.* J. J. Schofield, "Wooing by Wire," in *Lightning Flashes and Electric Dashes: A Volume of Choice Telegraphic Literature,*

Humor, Fun, Wit and Wisdom (New York, 1877), 164. All further page numbers will be included parenthetically in the text.

46. L. A. Churchill, "A Slight Mistake," in *Lightning Flashes and Electric Dashes: A Volume of Choice Telegraphic Literature, Humor, Fun, Wit and Wisdom* (New York, 1877), 74.

47. L. A. Churchill, "Playing with Fire," in *Lightning Flashes and Electric Dashes: A Volume of Choice Telegraphic Literature, Humor, Fun, Wit and Wisdom* (New York, 1877), 69. All further page numbers will be included parenthetically in the text.

5

From Phantom Image to Perfect Vision: Physiological Optics, Commercial Photography, and the Popularization of the Stereoscope

Laura Burd Schiavo

In 1853, about fifteen years after the introduction of the instrument, Charles Mansfield Ingleby of Trinity College, Cambridge wrote an essay about the stereoscope. Extolling it for its contribution to an understanding of binocular vision, Ingleby asked, "What dependence . . . can be placed upon the revelations of vision? What evidence do our eyes afford us of the solidity of seen objects?" His answer: The stereoscope demonstrated that eyes afford *no* dependable reports of objects, and that, in fact, "the objects of vision are but a mere phantasmagoria of the organ of sight."[1] Yet Ingleby's conclusion—that vision occurs independent of reality—was not the only one to which the stereoscope would lend itself. Indeed, a mere eight years later, Oliver Wendell Holmes implicitly praised the instrument precisely for its ability to reproduce sites accurately. In the second of three articles on photography and the stereoscope that he wrote anonymously for the *Atlantic Monthly,* Holmes recommended virtual tours of the United States and Europe, and concluded, "You may learn more, young American, of the difference between your civilization and that of the Old World by one look at this than from an average lyceum-lecture an hour long."[2] Taken together, Ingleby's and Holmes's ideas about the stereoscope illustrate competing discourses on vision at mid-century, one voiced by the developing field of physiological optics, and the other fostered by its applications in the burgeoning practice of commercial photography.

During the middle decades of the nineteenth century, the meaning of the stereoscope and stereo-viewing changed dramatically. Initially designed in 1838 to demonstrate a theory of vision, the stereoscope acquired new interpretations when it was commercialized during the 1850s and 1860s. Transforming the stereoscope into a popular amusement, photographers, retailers, and those in the optical trades not only promoted a vernacular form, the "parlor stereoscope," but also advanced a more positivist theory of vision that both relied upon and further reinforced the assumption that the subjects of sight were stable and that observation of those subjects led to accurate judgments. The masterful visual encounter attributed to the three-dimensional stereoscopic view, then, was not due to the medium's structure, but was rather the result of the inscription the instrument underwent as it became a consumer good.

In its original laboratory setting, the stereoscope raised fundamental questions about the status and reliability of vision. Remade into an instrument that delivered perfected, truthful representations, the parlor stereoscope was a far more marketable commodity. Yet the commercial motivations for the reinterpretation of the stereoscope should obscure neither the cultural field that supported this revision nor the viewing subject engendered by it. For a Euro-American culture desirous of instances of imperial display and invested in the creation and observation of the "world-as-exhibition," the stereoscope in its parlor form, complete with its attendant theory of vision, achieved a cultural significance that other optical toys—devices often clustered with the stereoscope in twentieth-century accounts of popular philosophical instruments—could never have satisfied.[3]

Decades of media studies and recent examinations of the materialities of communication conclude that media do not function as neutral systems for the storage and communication of information or opinion. In the language of Jean Baudrillard, media are "not *coefficients* but *effectors* of ideology."[4] The logical result of such insights has been a more prolonged and focused engagement with the systems, structures, and technological logics of communication and its tools. Yet the case of the stereoscope, in which one medium effected two ideologies—one scientific and the other commercial—cautions that scholarship, in the search for the specific structures of inscription, should not tread unscrupulously over variables of history, social usage, locations of power, and affiliated cultural production. The ideological "lessons" of the stereoscope changed dramatically as the device moved from scientific instrument to popular amusement, while the technological structure remained, to a great extent, intact. These "lessons" thus cannot be reduced to an intrinsic logic of the stereoscope and its operative form. Despite the relative consistency of the medium, the vernacular translation of the stereoscope did not reflect—even in simplified form—that of the scientific instrument from which it originated. Without a serious engagement with its uses and the conditions of its reappropriation, the cultural implications of the stereoscope, one of the most popular forms of nineteenth-century visual culture, remain woefully misunderstood.

Sir Charles Wheatstone first presented the stereoscope publicly in an address to the Royal Society on June 21, 1838.[5] For centuries philosophers and artists had remarked on the phenomenon of binocular vision: Given the space between the eyes, each eye is presented with a slightly different image, and yet we do not see doubly. It was not until

Wheatstone's demonstration, however, that the significance of binocularity to the perception of depth was proven.[6] Believing that the mind combined two dissimilar, monocular retinal images to create one, three-dimensional view, Wheatstone rhetorically inquired, "What would be the visual effect of simultaneously presenting to each eye, instead of the object itself, its projection on a plane surface as it appears to that eye?" and hypothesized "The observer will perceive a figure of three dimensions, the exact counterpart of the object from which the drawings were made."[7] When seen through the stereoscope, a pair of monocular, two-dimensional drawings merge into a three-dimensional solid. In Wheatstone's stereoscope, two mirrors anchored at right angles reflect drawings mounted on the instrument's arms, thus assisting the eyes in merging the images (figure 5.1). The instrument clearly demonstrated that the mental combination of dissimilar pictures led to the sensation of depth. Just how those two, flat fields were combined became a matter of debate, and the stereoscope lay at the heart of the controversy.

Philosophers, scientists, and artists from Euclid to Leonardo da Vinci had included optics, the study of light, among their many pursuits.[8] But it was not until roughly the

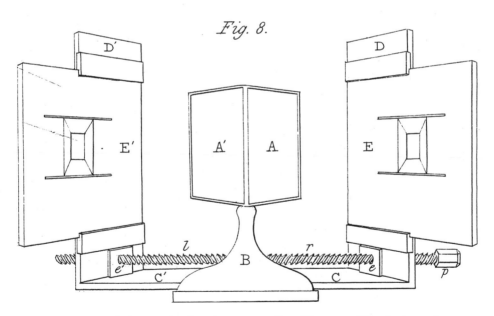

Figure 5.1   Drawing of Wheatstone's reflecting stereoscope. From Wheatstone, "Contributions to the physiology of vision—Part the first. On some remarkable, and hitherto unobserved, phenomena of binocular vision." *Philosophical Transactions of the Royal Society* 128 (1838).

middle decades of the nineteenth century that the study of vision emerged as an autonomous discipline, embracing characteristic instruments, methods, theories, and controversies.[9] By the 1850s that field, which came to be known as physiological optics and took as its subject "the study of the eye and its sensory capacities," was increasingly concerned with the phenomenon of the body rather than "the mechanics of light and optical transmission."[10] Owing largely to questions invoked by Wheatstone's stereoscope, the field was dominated by the investigation of depth perception, or how viewers understand the position of objects in relation to themselves. From 1850 to 1854, roughly a quarter of all writing on vision dealt with this question, a number that increased to nearly one third in the years following, during which time stereoscopic experiments provided central evidence.[11] These findings, which emphasized the vagaries and subjectivity of vision, destabilized earlier notions of a fixed and objective relationship between the exterior world and its transparent, visual realization, and demanded a reinterpretation of fundamental concepts about the body's role in perception.

For centuries natural philosophers and scientists had speculated about the relationships among objects in the visual field, images formed on the retina, and what is seen. Theories of vision dating from the Renaissance were based on a single, ideal eye, understood geometrical optics to be central to the study of vision and envisioned a direct correlation between the object and its retinal image. The eye was understood to function as a reliable, passive mechanism for recording the external world and transmitting its image. As for binocularity, the dominant model since the seventeenth century had resolved the question of single vision with two eyes by rationalizing that light rays emanating from objects stimulated corresponding points on the retina of each eye, in effect explaining the issue away by reasoning that the two eyes functioned as one.[12]

Wheatstone's experiments, much to the contrary, suggested that depth perception resulted from the mind's forced coalescence of dissimilar points. Stereoscopic experiments demonstrated that two unlike, two-dimensional projections were, by the actions of the body, made to fuse into one three-dimensional image, thus underscoring the body's involvement with vision. By inducing the illusion of solidity with only binocular *cues,* and prompting the experience of solidity where no depth actually existed, the stereoscope called into doubt the alleged subordination of vision to touch, an assumption predicated on the belief in a self-present world "out there." Creating a situation in which we "see" that which is not really there, the stereoscope insinuated an arbitrary relationship between stimulus and sensation. This notion challenged centuries of thought that had assumed a direct correspondence between objects and their retinal projections.

Jonathan Crary has argued that the classical archetype of the workings of the eye, in which a passive beholder observes the self-present world, gave way in the early nineteenth century to a new, subjective model of vision that located sight in the interaction between the body and the outside world. The indications of Wheatstone's stereoscope signal just such a transformation. Early stereoscopic experiments and consequent findings by Wheatstone and others point to the subjectivity of vision much in the way Crary describes. Despite some disagreement over its implications, the stereoscope and related experimental demonstrations had forever disrupted previous theories of vision.[13] By suggesting that vision could be manipulated into causing observers to see what was not really there, and by challenging the equivalence of the exterior world and the retinal image, the stereoscope suggested a model that ceased to suppress viewer's subjectivity in the manner of seventeenth- and eighteenth-century visual paradigms.[14] By introducing the body and its productive capacities into the story, the stereoscope contested the idea that vision could be represented geometrically—the basis for Renaissance perspective and the cornerstone of what Martin Jay has called the "scopic regimes of modernity."[15] It was no wonder that one scientist concluded in 1859 that Wheatstone's findings threatened a "complete overthrow of the theory of vision."[16]

By approaching the issue of binocular vision from the perspective of the body's awareness of depth—even in the absence of dimensionality—these experiments were phenomenological in nature. Wheatstone's pursuits diverged from a focus on the mechanism of the eye and attendant geometrical optics. In response to the issues that the stereoscope raised, Wheatstone, among other mid-nineteenth-century thinkers, developed theories that cast the body, rather than the world outside it, as an active producer of sensation. The phenomenological evidence provided by the stereoscope and other optical devices contributed significantly to the weakening of classical models of vision grounded in fixed spaces of representation, as perception became more and more a question of stimuli without reference to particular points in space. By the 1840s and 1850s, the science of vision had become both less geometrically oriented and less predictable. Scientists and philosophers expressed more and more interest in the irregularities of vision, including phantom images, optical illusions, and other manifestations of the subjectivity of vision.[17]

During this formative period, general-interest periodicals and popular works on science and industry mimicked the scholarly debate, duplicating the tenor of physiological optics if not taking sure positions on theories of visual perception. Fascinated with subjective vision, they too engaged with the growing realization of the disjunction between eye and object. In 1852 *Godey's Lady's Book,* an American magazine covering topics from fashion to sentimental fiction, reprinted an article from the *Illustrated London News* that emphasized the deceptive nature of vision and the capacity of the mind to produce illusion, describing stereoscopic visions as phantomlike, and declaring that "nothing can be more subject to deception than vision."[18] Extensive essays on the stereoscope in the popular works *Great Facts: Popular History and Description of the Most Remarkable Inventions during the Present Century* (1860) and *Fireside Philosophy or Familiar Talks about Common Things* (1861) described the physiology of binocular vision, mingling various theories of sight, and alluding to the arbitrariness between stimulus and sensation.[19] Popular texts like these provided a vernacular outlet for the discussion of the healthy mind's productive capacities and allowed for the indeterminacy of sight and the dissolution of ideal relationships between the exterior object and its visual counterpart.

Commercial photographers soon recognized the marketability of a merger between photographic technologies and an instrument that could turn the two-dimensional photograph into a three-dimensional tableau. By 1855 the *American Journal of Science* could report the success of this consolidation, crediting the "remarkable discoveries" of Wheatstone and others with being "among the most important contributions ever made to the science of physiological optics," and continuing, "Nor has the knowledge of these beautiful and curious results been confined to the circle of scientific inquirers. The diffusive intelligence of the age has converted the stereoscope into a popular source of instructive recreation."[20] Mid-century photographic histories made evident the ease with which photographers incorporated the stereoscope into the history of their art and trade and, conversely, denigrated Wheatstone's original stereo-views. According to these narratives, the limitations of hand-drawn diagrams (in contrast to photographs) severely curbed the instrument's utility. As Nathaniel Burgess wrote in his *Photograph Manual; A Practical Treatise, Containing the Cartes De Visite Process, and the Method of Taking Stereoscopic Pictures* (1860):

> this truly wonderful result (of the stereoscope) cannot be obtained unless the drawings are exact copies of nature, even more exact than the human hand can execute. We are therefore obliged to call in the aid of Photography, without which Stereoscope pictures would have never obtained that perfection which we see now exhibited.[21]

Photographic histories described a natural and mutually advantageous relationship between the technologies of stereoscopy and photography, containing stereoscopy as a novel and promising division within the field of photography. Without photography, the stereoscope would have been relegated to either "a retired and passive, though honourable, existence in the world of science," or remained a "mere toy," similar to the status of a faddish curiosity shared by such optical devices as the phenakistoscope and kaleidoscope.[22] Even those aware of the independent origins of the two technologies commented on the fitness of their association. James Ellis argued in the tellingly titled *Progress of Photography, Collodion, the Stereoscope* (1856) that "although they have no natural relationship," by their "rightful marriage" the stereoscope and photography "are indissolubly joined together for their mutual advantage." Ellis described the interdependence of the two technologies and identified "the efficacy of each for the dignification of the other," calling the ten-year lag between the stereoscope's introduction and stereo-photography "among the curiosities of the history of science."[23] Narratives like Ellis's and Burgess's reveal the triumph of a photographic prerogative in the popular understanding of the stereoscope by the 1860s.

A brief glance back to Wheatstone's device reveals the extent of this appropriation. The decade of stereoscopy without photography that seemed so incredible to Ellis was not merely an accident of technological timing. Wheatstone had purposefully used eleven pairs of simple line-drawn shapes and figures including cubes and cones with his stereoscope, contriving each so that it represented the two projections of an object seen from two points of sight (figure 5.2). Their simplicity was both intentional and substantive. Only the most elaborate of these drawings, a pair of shaded arches, included any conventional allusions to depth. Because Wheatstone's demonstration sought to display the visual circumstances that contributed to the brain registering dimensionality, he had to isolate the variable of binocular disparity. Only by erasing other visual cues of solidity and space could the perceptual factors of sight be displayed. Pictorial elaborations of depth (such as the conventions of shading or atmospheric effects) would have introduced the possibility that it was they, rather than the perspectival differences in the images, that created the effect. Photographic substitutions would only have confused, not improved, Wheatstone's demonstration.[24] Rather than interpreting the stereoscope in the absence of photography as inherently unfinished, then, we might more accurately highlight the *commercial* inviolability of Wheatstone's stereograms.[25]

Where nineteenth-century accounts overstated the shared purpose of stereoscopy and photography, Jonathan Crary's more recent reevaluation overcompensates in the other direction. In *Techniques of the Observer* Crary contrasts two models of the observer and two corresponding theories of vision: The classical model, symbolized for Crary by the camera obscura, imposed a notion of vision as stable, transcendent, and passive, and a rational viewing subject similarly described; while a new, modern, subjective model of vision

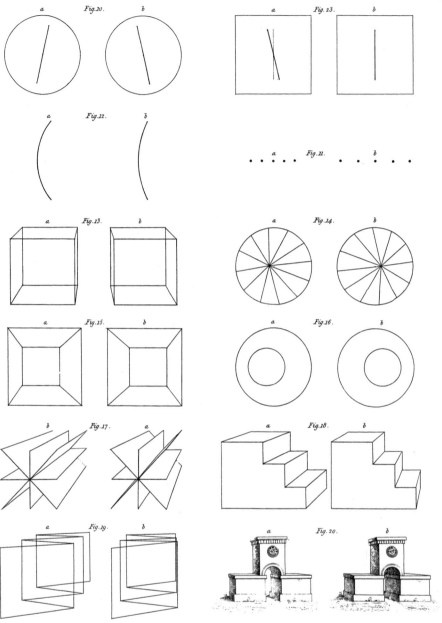

J. Basire sc.

Figure 5.2   Stereoscopic drawings. From Wheatstone, "Contributions to the physiology of vision—Part the first. On some remarkable, and hitherto unobserved, phenomena of binocular vision." *Philosophical Transactions of the Royal Society* 128 (1838).

located sight in the interaction between the body and the outside world, where clear distinctions between the interior and the exterior were muddied, and for which the stereoscope is the emblematic device. In this shift, Crary claims, borrowing from Jean Baudrillard, "the problem of mimesis disappears."[26] Given the centrality of this paradigmatic transformation to his work, Crary is rhetorically compelled to disentangle the subjective vision of the stereoscope from the complications of what, in his duality, are the contending suppositions of photography (with its clear roots in the camera obscura). Thus, owing to the nature of his argument and its academic distinctions, Crary displaces photography as a dominant perceptive model, deemphasizes the centrality of photography to stereoscopy, and overlooks the ways in which the distinctions between the two media were obscured in the practice of commercial stereo-photography.[27]

In an effort to boost their still inchoate business, commercial photographers relied heavily on what Anthony Hamber has called "idiosyncratic photographic print formats," one of the most successful of which was stereo-photography.[28] Indeed, with the close of the daguerreotype era and the introduction of reproducible photography on paper (thanks to the technological advances of collodion glass plate negatives and albumen prints), stereographs were one of the most significant formats by which the general public came in contact with photography.[29] The integration of stereoscopic hardware and stereographic software into commercial photography was fully entrenched by the late 1850s, when scores of photographers were producing in stereo, and photographs, stereographs, and stereoscopes were sold side by side in photographic studios, optical shops, and retail outlets, and by mail order[30] (figure 5.3).

In formulating a vernacular, camera-based iteration of the stereoscope's significance and purpose, photographers and cultural critics proposed a theory of vision that held fast to Crary's classical model of vision with its suppression of subjectivity and its grounding in a faith in human optics as an accurate reflection of the external world. In similarly worded catalogs, trade publications, and editorials, they promised eminently reproducible, truthful depictions of lifelike solidity. This was a far cry from theories being presented in physiological optics about the productive nature of vision. With the conflation of two virtually concurrent technologies of vision—stereoscope and photography—the meaning of stereoscopic viewing was recast in the public imagination. Reconfigured as constituents of commercial photography, stereoscopic views were removed from the field of phenomenological inquiry and contained in the language of fine arts as models of mimetic representation.

Figure 5.3   D. Appleton & Co., Stereoscopic Emporium, 346–348 Broadway, New York. From stereograph by D. Appleton & Co., circa 1860. Courtesy of National Museum of American History, Smithsonian Institution (catalog # 71.59.4).

Photographers and cultural critics were encouraged in their reformulation of the meaning of the stereoscope in large part by the design of a new stereoscope intended for popular consumption. At London's Crystal Palace Exhibition in 1851 Sir David Brewster introduced such a stereoscope—smaller, more compact, and more commercially viable than Wheatstone's model.[31] Brewster's lenticular stereoscope, so called because it substituted lenses for the mirrors of Wheatstone's reflecting model, was roughly eight inches long and five inches wide. While he designed more than one pattern, Brewster's prototypical model is a pyramidal body of tin or wood, with two eye-pieces made of half-lenses mounted in the front panel and separated by about two and half inches (a distance equal to the space between the eyes).[32] Far more convenient and less awkward than Wheatstone's, the Brewster stereoscope refined the instrument's market appeal. So too did his choice of venue: By introducing his version of the instrument at the Crystal Palace, Brewster ensured that his stereoscope became part of the display of goods and manufactured products meant to demonstrate the British Empire's civilization and progress. The exhibition, in the words of one observer, would display the spirit of the age and "advance our National Taste."[33]

Owing to their protective glass covering and the ensuing glare, daguerreotypes had been difficult to view in Wheatstone's open, skeletal instrument. The lenticular model, rather than merely accommodating them, was designed with the display of daguerreotypes in mind, thus solidifying the relationship between the stereoscope and photography.[34] What's more, the boxlike framework enhanced the stereoscope's commercial appeal and function. Despite its smaller frame, the enclosed form had more surface area, which allowed for greater ornamentation. As early as the 1850s, manufacturers were crafting Brewster stereoscopes in fine woods and decorative papier-mâché, many inlaid with ornamental patterns and mother-of-pearl, or covered in morocco leather or cloth (figure 5.4). Moreover, the small size and the enclosed format mandated that only mounted images of a certain size could be used, forcing a standardization of stereograph cards. Sixth-plate stereo-daguerreotypes and $3\frac{1}{2}'' \times 7''$ stereographs met these needs and became stock industry format, introducing interchangeability to the stereo industry. Brewster's lenticular model, in its compact size, its potential for ornament, and its newly incorporated standardized software, became a commercially viable emblem of parlor culture.

The lenticular stereoscope also downplayed its operative mechanism in ways that helped deny what had been considered the illusionistic nature of binocular vision as explored in physiological optics. Two half-lenses mounted in the front panel made up the sum total of the apparatus and were effectively hidden by the enclosed form—a considerable variation from the long skeletal arms and large mirrors of Wheatstone's reflecting model. Once slid into the stereoscope, the card stereograph vanished as well. Concealed components produced an instrument that resembled opera glasses more than it did laboratory equipment, a design that anticipated an audience unconcerned with the technicalities of optical convergence.

Borrowing from Karl Marx and Theodor Adorno, Jonathan Crary has argued that such transformations approximate the potentially "phantasmagoric" nature of optical devices. An apparatus openly displayed (as was the case with Wheatstone's reflecting stereoscope) fractures the observer's attention from the resulting effects or surfaces. In the manner of commodity fetishism, when those component parts are successfully hidden (as they are by Brewster's lenticular stereoscope) *only* the effects surface. In this way, Brewster's model masked the phenomenological event and physicality of binocular vision—its productive, human labor—presenting optical illusion as commodity. In Adorno's words,

Figure 5.4   Papier-mâché stereoscope with mother-of-pearl inlay (England), circa 1850. Courtesy of Strong Museum, Rochester, New York © 2001.

aesthetic appearance becomes a function of the character of the commodity. As a commodity, it purveys illusions. The absolute reality of the unreal is nothing but the reality of a phenomenon that not only strives unceasingly to spirit away its own origins in human labor, but also, inseparably from this process, and in thrall to exchange value, assiduously emphasizes its use value, stressing that this is its authentic reality, that it is "no imitation."[35]

By hiding the physiological roots of the stereoscopic image, Brewster's stereoscope omitted the productive or phenomenologic nature of the device, and of vision itself, making the stereoscope nothing more than a tool for the enhancement of mimetic representation. With this concentrated shift to mimetic codes, the stereoscope was wholly reformulated into a commodity form—a popular amusement and a purveyor of unmediated truths, a precursor of cinematic diegesis. Despite Marx's useful terminology, however, the results were far from "phantasmagoric" in the sense of ghostly illusions and magic lantern projections. The illusion here was one of positivism and avowed transparency. Conceptually, the new stereoscope rested on a classical notion of transparent, unmediated representation.

Despite its eminently commercial appeal as a commodity form, the popular stereoscope gained legitimacy in part through an imprecise yet insistent association with scientific pursuit. Stereo catalogs and general-interest periodicals never hesitated to remind their readers that the origins of the stereoscope lay in the laboratory with the study of vision. An early catalog of McAllister & Brother of Philadelphia, one of the first optical firms to sell stereoscopes in the United States, described the instrument as "the result of the investigations on the subject of Binocular vision, which have been pursued for some years past by eminent scientific men in Europe."[36] Illustrated catalogs of optical instruments presented drawings of stereoscopes next to those of microscopes and telescopes, further stressing the scientific roots. Text after text located the etymology of Wheatstone's 1838 neologism in the Greek for "solid" and "I see."[37] One handbook of parlor games, linking the classicism of the declension to erudition, commented that the stereoscope, "one of the newest and most interesting optical surprises invented . . . like many other instruments . . . is indulged with a very hard name."[38]

In what became a fairly typical interpretive strategy (that, not coincidentally, coincided with the origins of the stereoscope), writers in the popular press inevitably began their discussions of the instrument with investigation of the structure of the eye. Having alluded to the stereoscope's origins as a tool with which the workings of the eyes had been studied by men of science, popular sources portrayed the device as a mechanized model by which lay people might also comprehend and appreciate the workings of vision. Articles often suggested simple experiments by which to confirm the fact of binocular vision.[39] Photography manuals, popular texts on science, and general interest and trade publications all emphasized the double vision of the stereo format and its resemblance to the two eyes. They based this equivalence on two analogous relationships: that between the two lenses of the stereoscopic camera and the two lenses of the eyes; and that of the two retinal images and the double images of the stereograph. This interpretive program located the stereoscope's value not only in the consequent dimensionality of its representations but in the mechanics of the instrument and its views.

According to Nathaniel Burgess's 1862 photography manual,

> The philosophy of the Stereoscope has been explained as the true philosophy of vision; that as we view all objects through the medium of two eyes, the two impressions are made upon the retinae of the two eyes, precisely in the same manner as the Stereoscopic impression is made by the camera obscura. Each eye has a distinct reflection of all objects brought within its view, and there is no doubt of the fact, that there are represented the two pictures, as accurately as seen in the Stereoscopic production; . . . This is called binocular vision, or seeing two objects at once with the two eyes.[40]

Here Burgess, like many others writing popularly and for the photographic trades, seemingly resolved the complexities of binocular vision through a declaration of its application to pictorial representation. This reductive parallelism of lens for eye revived an earlier notion that had naturalized photographic vision—the conflation of the camera obscura (and later the camera lens) with the lens of the eye. This model, rooted in a geometrical optics and based on the single-eye model, figured the eye as the ideal optical instrument and the standard by which all other technologies of vision were judged.[41] Given the physical fact of binocularity, this model was enthusiastically adapted to, and made more potent by, the double lenses of the stereoscopic camera and the double images of the stereograph.[42] Burgess's direct reference to the camera obscura underscores the centrality of the photographic/ocular metaphor to the stereoscope's vernacular interpretation and indicates the successful naturalization of stereoscopic vision.

The complexities of binocularity that had disrupted older theories of vision in physiological optics gave little pause to popular writers and photographers whose vernacular spin naturalized stereoscopic vision as a faithful imitation of human perception. As attested to in *The Magician's Own Book* (1857) the stereoscope was "an imitation of the powers of the eyes":

That is to say, if it were possible to be behind the retina of each eye, and draw the two pictures of any object seen by our eyes, those pictures put into the stereoscope, would reproduce the solidity from which they were drawn.[43]

Such claims articulated a notion of stereographic representations as transparent, reflective, wholly reliable transcriptions of retinal images, themselves unfailing equivalents to the external world they signified. They reasserted the equivalence between the exterior world and its counterpart in the eye that Wheatstone's laboratory instrument had called into question, emptying the stereoscope of the very notion of the body's productivity and the phenomenological interest so central to physiological optics. They repudiated the more complicated—and less marketable—explanations of the relationship between the viewing subject and the visual field, replacing it with the one-to-one correlation between the world and its stereoscopic recreation. No longer an instrument for phenomenological interrogation and the study of the vagaries of vision, the stereoscope had become a tool for furnishing visual truths.

Popular discussions added one more level to the representational accomplishments of the stereoscope and its views. In its avowed re-creation of retinal images, the stereoscope became an archetype of what photographers and cultural critics of the day described as "perfect vision." Formerly the centerpiece for a discussion of the disjunction between an object and its visual representation, the stereoscope came to symbolize their perfect correlation. An article originally appearing in the *Illustrated London News* in 1852 and reprinted in the *Photographic Art-Journal* that same year provides an early articulation of the concept of "perfect vision" in connection with the stereoscope. Having described the relationship between photography and stereoscopy, the article concludes, "There is, therefore, no further room for doubt as to the need for two eyes; we have taken by the daguerreotype the very picture from each, and have made them tell their secret. Our double vision is but perfect vision."[44] This formula portends an end to skepticism not only about the status of stereoscopic representation but about the precision of vision. It affirms the truth-telling nature of the photograph (in this case a daguerreotype), and posits that the very doubleness of the stereograph "proves" the need for two eyes. The fusion of stereoscopy and photography, the argument tautologically alleges, simultaneously proves the flawlessness of nature and replicates the world faithfully for our viewing pleasure. The harmony, indeed the natural compatibility, of the stereoscope and the camera obscura professed here belies the argument that would consider the stereoscope without the double photographs that were its partners in meaning.

But the stereoscopic camera was more than an enhancement of its monocular counterpart. Employing the distinctions of American semiotician Charles Sanders Peirce, photography bears more than an iconic resemblance with the objects it represents (as would be the case with a painting or other graphic representations). Rather, an effect of light radiating from an object, the photograph registers the very physicality of the object, is in fact caused by that object, and thus may be classified as indexical in Peirce's taxonomy.[45] With the doubling of vision in the stereoscope, the stereograph made a qualitative leap in photographic claims to indexicality. By alleging a direct correspondence between retinal and stereographic images, popular writers utilized binocularity to new advantage: The stereoscope and stereograph were credited with recreating natural viewing conditions, with reconstituting the very visual circumstances that would have presented themselves to the eyes had the viewer witnessed the scene herself. The stereograph thus signifies in a manner beyond that of the indexical sign, becoming instead the actual visual equivalent of the object or objects signified. The stereograph seen through the stereoscope was alleged to be the very same as the scene viewed in person. Photographer Robert Hunt put it simply in the pages of the *Art Journal* in 1856, applauding the stereoscopic format for enabling the viewer to "see things as they are in nature."[46] Likewise, an 1871 article in a photography trade journal placed a presumptive condition on the positive reception of stereographs: "Recreation" and "advantage" would only be accomplished with stereoscopic photography "as long as it is conceded that the photographic views of most objects are seen more vividly and naturally in this than in any other manner."[47]

This notion of visual equivalence was clearly voiced in the introduction to the first of a two-volume set, *The Stereoscopic Magazine: A Gallery of Landscape Scenery, Architecture, Antiquities, and Natural History,* published by Lovell Reeve in London in 1858. Predicated on a synergistic relationship between "the magic power of the Stereoscope" and the "truthfulness of the camera-obscura," the work presented stereo-photographs and lengthy textual descriptions of sites in Western Europe including monuments, museums, cathedrals, castles, gardens, and natural wonders.[48] An avowed parallelism between technology and the body thus informed Reeve's assumptions.

> The Stereoscope, with its *two* lenticular arrangements, may be supposed, for the moment, to represent the *two eyes* in *one head*. . . . Without entering more deeply into this question, it will be sufficient to state that the pair of pictures upon the stereoscopic slide are, as nearly as possible, such a pair of pictures as are formed upon the retinae of the eyes, supposing they looked upon the scene or subject from which *it* had been copied, and the Stereoscope is nothing more than an instrument by which those pictures are directed, respectively, to the right and to the left eye of the observer.[49]

This project exemplifies the scientistic and positivist claims and suppositions that became the hallmark of the stereoscope and stereoscopic views in the 1850s and 1860s and that helped to determine its later history as a popular amusement and educational aid. When, in 1862, the Bierstadt brothers of New Hampshire published their own stereographically illustrated tourist companion, *Stereoscopic Views among the Hills of New Hampshire,* they too alleged that "The Stereoscope presents to the eye all the objects in solid relief, as perfectly as if the landscape itself were spread out before it."[50] The book included a pair of lenses attached to a flap on the inside cover of the book for convenient viewing of the illustrations (figure 5.5).

Productions like *The Stereoscopic Magazine* and *Stereoscopic Views Among the Hills* realized the promises extended earlier in the decade. In 1853 the *Art Journal* wrote that through the stereoscope "on the tables of our own drawing-rooms may we examine at our leisure, those far-distant scenes in which we are interested, without the toil of travel."[51] The *Home Journal,* a New York newspaper, similarly declared in 1859,

> Since the discovery of photography, the world abroad, as well as at home, has been brought, as it were, into our very houses; and with the aid of stereoscope, we can visit, with our sight, all the memorable places on the globe; we can gaze on the most celebrated works of art, and look on all the distinguished men of the age.[52]

By 1871 such dreams for the stereoscope assumed an even more practical and commercial design. If a traveler wished to journey from New York to Niagara Falls, one article suggested, he might stereoscopically preview various itineraries at ticket offices, where views would "no doubt tempt him to go one way, return the other, and seduce (him) into a second trip to test the third."[53] Either in the re-creation of views already seen, or in the presentation of scenes not yet beheld, the stereoscopic view was the perfect souvenir, enacting the scene by stimulating the eyes precisely as they would have been had one been there oneself.

A correspondent, identified only as G.A.S., sent a letter to *Notes and Queries* in 1868 in which he inquired about a newspaper clipping dated twenty years earlier. The article to which he referred described, but did not name, an invention by a Professor Wheatstone. G.A.S. asked the editor whether the instrument specified might be the derivation of the popular parlor stereoscope.[54] Indeed it was. As G.A.S.'s letter indicates, by the late 1860s the stereoscope's origins in the laboratory of Sir Charles Wheatstone had, popularly speaking, been largely erased. In forgetting the reflecting stereoscope and its position in physiological optics it is likely that the public forgot as well much of what the

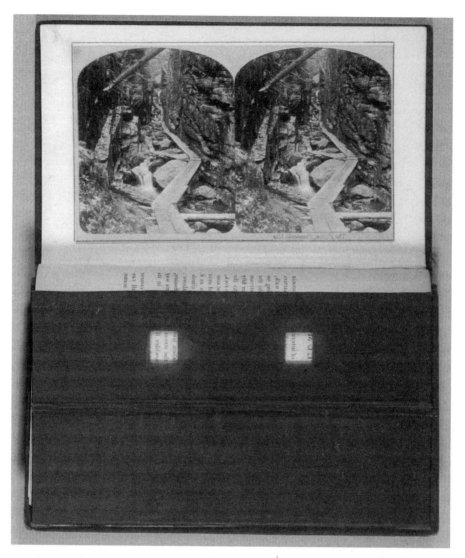

Figure 5.5   Charles Bierstadt and Edward Bierstadt, *Gems of American Scenery, Consisting of Stereoscopic Views among the White Mountains with Descriptive Text* (1875) (Niagara Falls, New York). Courtesy of American Antiquarian Society.

instrument had proposed about the subjective nature of vision and the productive role of the body in perceiving the world. In the roughly two decades between the invention of the stereoscope and G.A.S.'s letter, commercial photography successfully absorbed the stereoscope through persistent and repetitive writing on the instrument that appeared not only in trade literature and advertising materials but in parlor manuals and popular periodicals as well. These texts insisted on a world eminently and objectively exhibitable, equivalently revealed by their stereo-photographic representations, souvenirs to be quickly swallowed up by what *Anthony's Photographic Bulletin* gleefully termed in 1870 the "vortex of popular consumption."[55]

The commercial motivations for this reinterpretation were powerful. Yet it is also crucial to consider the more amorphous cultural field that sustained such a transformation in interpretation. As Nicholas Wade and Michael Swanston maintain, visual perception might most simply be defined as the process of "localising objects accurately, so that we can step over them, reach for them, avoid them or orient appropriately to them."[56] The parlor stereoscope granted its viewers unequalled visual access. Where Wheatstone's stereoscope had raised questions about the relationships between objects and viewers, and about the truth of appearances and the knowability of the world more generally, the popular stereoscope, incessantly defined within the world of photography and popular amusement after its commercial introduction in 1851, stabilized and solidified those very relationships. The parlor stereoscope and its views enacted a confidence in vision and in the transparency between the object and its representation. They functioned as material evidence—consumable and repeatable signs—of abstract dreams of nineteenth-century imperial vision.

A scientific instrument devised to demonstrate one concept of vision became a popular device whose interpretation and habitual use asserted another. This narrative contributes additional evidence to the more than amply documented argument that scientific instruments, and in some cases the popular media that they may become, can no longer be characterized as neutral, transparent technologies through which empirical truths are revealed. The process of stereoscopic inscription fixed those views with meaning; the messages quite clearly contained traces of the medium.[57] But the fact that shifting ideological and philosophical meanings were not wholly constrained by or contained in the structure of the medium makes us turn for explanation to culture. Based on the assumption that human optics accurately reflected external reality and that photography in concert with stereoscopy could replicate the conditions of vision, stereo-photographic depictions were widely reproduced and disseminated for the visual pleasure and intellectual refinement of an acquisitive nineteenth-century audience. They did so at a formative moment in the 1850s and 1860s, decades before the introduction of the halftone most often credited with incorporating photography into mass-illustrated materials and enhancing "the illusion of seeing an actual scene, or receiving an objective record of such a scene."[58] Foreign and domestic locations, home towns, depictions of current events, celebrities, scientific specimens, worldly treasures, fine and decorative arts, ethnic and racial groups, and exotic peoples in foreign locales were all situated in viewers' hands, accessible for their viewing pleasure, lying quite literally in their uncomplicated lines of sight.

Notes

1. C. Mansfield Ingleby, "The Stereoscope Considered in Relation to the Philosophy of Binocular Vision," (London: Walton & Maberly, 1853), 24.

2. Oliver Wendell Holmes, "Sun-Painting and Sun-Sculpture; with a Stereoscopic Trip across the Atlantic," *Atlantic Monthly* 8, no. 45 (1861): 23.

3. The idea of "world-as-exhibition" is indebted to the work of Robert Rydell on international exhibitions and specifically to Timothy Mitchell's reading of the British display of Egypt. Mitchell describes how "representations" set up "the world as a picture . . . as an object on display, to be viewed, experienced and investigated." See Timothy Mitchell, *Colonising Egypt* (Cambridge: Cambridge University Press, 1988), 6, 7–13. With regard to optical toys I refer to the kaleidoscope and devices based on the persistence of vision, including the thaumatrope, the phenakistiscope, and the zoetrope. Texts that have treated these devices as a group include Jonathan Crary, *Techniques of the Observer: On Vision and Modernity in the Nineteenth Century* (Cambridge: MIT Press, 1990); Linda Williams, "Corporealized Observers: Visual Pornographies and the 'Carnal Density of Vision,'" in *Fugitive Images: From Photography to Video,* ed. Patrice Petro (Bloomington: Indiana University Press, 1995); and Barbara Stafford, *Artful Science. Enlightenment Entertainment and the Eclipse of Visual Education* (Cambridge: MIT Press, 1994), 67–70.

4. Jean Baudrillard, quoted in Geoffrey Winthrop-Young and Michael Wutz's Introduction to Friedrich A. Kittler, *Gramophone, Film, Typewriter* (Stanford: Stanford University Press, 1999), xv.

5. At the meeting Wheatstone delivered a paper, "On Binocular Vision; and on the stereoscope, an instrument for illustrating this phenomenon." For the written account, see "Contributions to the physiology of vision—Part the first. On some remarkable, and hitherto unobserved, phenomena of binocular vision," *Philosophical Transaction of the Royal Society,* 1838, reprinted in *Brewster and Wheatstone on Vision,* ed. Nicholas Wade (London: Academic Press, 1983), 65–93.

6. Wade, *Brewster and Wheatstone on Vision,* 14.

7. "Contributions—Part the First," in Wade, *Brewster and Wheatstone on Vision* 67, 69.

8. Literally defined as the geometrical properties of light, optics had come to include the science of vision as it was perceived to be governed by such laws.

9. For this development, see Steven R. Turner, *In the Eye's Mind: Vision and the Hemlholtz-Hering Controversy* (Princeton: Princeton University Press, 1994), 11.

10. Crary, *Techniques of the Observer* 88, 70.

11. Turner, *In the Eye's Mind* 17, appendix, table 2.

12. Nicholas J. Wade and Michael Swanston, *Visual Perception: An Introduction* (London: Routledge, 1991), 123. For a contemporary account, see Joseph Ellis, *Progress of Photography, Collodion, the Stereoscope* (London: Bell & Daldy, 1856), 36–45.

13. Conclusions drawn from the stereoscope did not go unchallenged or undisputed. Some natural philosophers, including Sir David Brewster, inventor of a lens-based stereoscope that became the basis for the popular model, resisted explanations that troubled earlier notions of a single eye. For Brewster, the stereoscope's "lesson" lay not in the unique contribution of dissimilar retinal images to depth perception, but in its affirmation of the centrality of vision to the perception of space. See Turner, *In the Eye's Mind,* 14–16.

14. This summary and interpretation is indebted to Jonathan Crary's reading of similar materials in Crary, *Techniques of the Observer,* chapters 3 and 4.

15. Martin Jay, "Scopic Regimes of Modernity," in *Vision and Visuality,* ed. Hal Foster (New York: The New Press, 1988).

16. A. W. Volkmann, as quoted in Turner, *In the Eye's Mind* 18. Wheatstone's conclusions predated Herman von Helmholtz's publication of *Physiological Optics* and his argument about the empirical nature of vision. See Wade, ed., *Brewster and Wheatstone on Vision,* 27.

17. Britt Salvesen, "Selling Sight. Stereoscopy in Mid-Victorian Britain," Dissertation, University of Chicago, 1997, 51.

18. "The Stereoscope," *Godey's,* October 1852, 342. This article had already been reprinted by the *Photographic Art-Journal,* one of many examples of redundancy of popular discussions of the stereoscope. In many texts, original sources were not acknowledged.

19. Frederick C. Bakewell, preface to *Great Facts: Popular History and Description of the Most Remarkable Inventions During the Present Century,* ed. Frederick C. Bakewell (New York: D. Appleton and Co., 1860); *Fireside Philosophy or Familiar Talks About Common Things* (New York: W.A. Townsend & Co., 1861), 297.

20. W. B. Rogers, "Observation on Binocular Vision," *American Journal of Science* 20 (second series), no. 58 (1855): 86.

21. N. G. Burgess, *The Photograph Manual; a Practical Treatise, Containing the Cartes De Visite Process, and the Method of Taking Stereoscopic Pictures* (New York: D. Appleton & Co., 1862), 196.

22. Texts that discuss stereoscopy as a division of photography include Ellis, *Progress of Photography, Collodion, the Stereoscope,* 51–52; Burgess, *The Photograph Manual,* 193; and John Henderson, "Photography and the Stereoscope," *Anthony's Photographic Bulletin* 2, no. 6 (1871): 162. Henderson's article in *Anthony's Photographic Bulletin* claimed that "the stereoscope received its impetus mainly, if not entirely, owing to its early connection with the infant art-science of photography. . . ." Regarding the status of the stereoscope in the absence of photography, see M. P. Simons, "A Plea for the Little Stereoscope and the Stereoscopic Portrait," *Anthony's Photographic Bulletin* 3, no. 3 (1872), 469. Still others reversed the relationship, constituting the stereoscope as the benefactor to photography which derived "its chief glory" only in the exercise of stereoscopic principles. According to Charles Himes, "the stereoscope multiplies wonderfully the applications of photography." Charles F. Himes, *Leaf Prints or Glimpses at Photography* (Philadelphia: Benerman & Wilson, 1868), 15.

23. Ellis, "Progress of Photography, Collodion, the Stereoscope," 50.

24. Wheatstone, "Contributions—Part the First," in Wade, *Brewster and Wheatstone on Vision,* 72. According to Wheatstone, "Had either shading or colouring been introduced it might be supposed that the effect was wholly or in part due to these circumstances, whereas by leaving them out of consideration no room is left to doubt that the entire effect of relief is owing to the simultaneous perception of the two monocular projections, one of each retina."

25. There is some indication that Wheatstone tried unsuccessfully to stimulate interest in pairing his stereoscope with daguerreotypes in the 1840s. See Ellis, *Progress of Photography, Collodion, the Stereoscope,* 50; "Lenticular Stereoscope," *Journal of the Franklin Institute,* July 1852; and Steven F. Joseph, "Wheatstone and Fenton: A Vision Shared," *History of Photography* 9, no. 4 (1985): 305.

26. Crary, *Techniques of the Observer,* 12.

27. In Crary's words, the stereoscope "*preceded* the invention of photography and *in no way required* photographic procedures or even the development of mass production techniques" (17). While this is true, it is misleading if left at that. Similarly, Crary explains that he concentrates on "the period during which the technical and theoretical principles of the stereoscope were developed," thus bracketing out of consideration all evidence of commercial practice and diffusion—the "issue of (the stereoscope's) effects once it was distributed throughout a sociocultural field" (118). Both Britt Salvesen and David Phillips pose similar criticisms of Crary's treatment of the position of photography in its relationship to the stereoscope. Salvesen, "Selling Sight," 69–70; David Phillips, "Modern Vision," *Oxford Art Journal* 16, no. 1 (1993): 135–137.

28. Anthony J. Hamber, *"A Higher Branch of Art": Photographing the Fine Arts in England, 1839–1880,* ed. Helene E. Roberts and Brent Maddox, vol. 4, *Documenting the Image* (Amsterdam: Gordon and Breach, 1996), 6.

29. Alan Trachtenberg, "Photography: The Emergence of a Keyword," in *Photography in Nineteenth-Century America,* ed. Martha A. Sandweiss (Forth Worth: Amon Carter Museum, 1991), 38.

30. D. Appleton & Co., one of the first U.S. firms to import views from Europe for sale in the United States (beginning in 1851), began domestic production as early as 1856, sold stereo views and stereoscopes, and developed innovative marketing strategies. Peter Wing, *Stereoscopes: The First One Hundred Years* (Nashua, N.H.: Transition Publishing, 1996), 53. The New York firm of E. and H. T. Anthony, which would become one of the largest stereo manufacturers, retailers, and wholesalers in the United States, sold stereoscopes, stereographs, large format photographs, frames, albums, and chromolithographs by mail-order catalogs and in their newly named Stereo-scopic Emporium, which opened at 501 Broadway in May 1860. See William and Estelle Marder, *Anthony: The Man, the Company, the Cameras* (Amesbury, Mass.: Pine Ridge Publishing Co,, 1982), 114. The number of views published is difficult to determine. William Darrah estimates the total at six million different stereo views published in the United States alone before the turn of the century. William C. Darrah, *The World of Stereographs* (Nashville: Land Yacht Press, 1997), 6.

31. David Brewster was no stranger to business, having marketed another optical device as a popular amusement over twenty-five years earlier. Out of his investigations of the polarization of light in the 1810s, Brewster developed the kaleidoscope, an instant sensation when marketed as "a popular instrument for the purposes of rational amusement." (David Brewster, *A Treatise on the Kaleidoscope*, 1819), as quoted in Wade, *Brewster and Wheatstone on Vision*, 204. According to his own writings, only the mismanagement of the patent prevented Brewster from realizing huge profits; Margaret M. Gordon, *The Home Life of Sir David Brewster*, quoted in Salvesen, "Selling Sight," 44–45.

32. For a description of Brewster's and other European and American models, see Wing, *Stereo-scopes: The First One Hundred Years*. The Brewster models are described on 3–8.

33. "The Art-Journal Illustrated Catalogue of the Industry of All Nations" (London: Art Journal, 1851), 1. This is one among many publications so describing London's Great Exhibition.

34. During the 1850s competing claims to the invention of the stereoscope became the subject of vituperative debate between Wheatstone, Brewster, and their supporters and detractors in the photographic and scientific press. While there is little doubt as to Wheatstone's invention of the stereoscope, Brewster's contribution in developing an instrument designed for the market was significant and assertions about his innovation are not without basis. *Scientific American* noted in 1853, "We believe it was Prof. Wheatstone, of London, who first made the discovery of the stere-oscope, which was afterward greatly improved by Sir David Brewster, and by him (Brewster) first applied to produce binocular vision with daguerreotype pictures." "Mascher's Stereoscope," *Scientific American* 8, no. 37 (1853): 292. *The Stereoscopic Magazine* similarly described Wheatstone as responsible for the development of the "theory on which stereoscopic effects depend" and the construction of instrument, but credited Brewster with the creation of the "simpler," popular form. *The Stereoscopic Magazine: A Gallery of Landscape Scenery, Architecture, Antiquities, and Natural History*, vol. 1 (London: Lovell Reeve, 1858), 5.

35. This reading of the "phantasmagoric" nature of the popular stereoscope is indebted to Crary's argument in relation to Emile Reynaud's modification of his praxinoscope into a projection

device he named the Praxinoscope-Théâtre in the late nineteenth century. Jonathan Crary, *Suspensions of Perception Attention, Spectacle, and Modern Culture* (Cambridge: MIT Press, 1999), 259–265. Theodor Adorno, *In Search of Wagner* (NLB: London, 1981), 90, quoted in Crary, page 261, n. 270.

36. McAllister & Brother, "Stereoscopes and Stereoscope Views" (Philadelphia, 1858): 1.

37. Hunt, "The Stereoscope," 274; F. Moigno, "The Stereoscope," *Photographic Art-Journal* 5, no. 3 (1853): 137; McAllister & Brother, "Stereoscopes and Stereoscope Views" (Philadelphia, 1858), 1; and *Fireside Philosophy or Familiar Talks About Common Things* (New York: W. A. Townsend & Company, 1861), 297.

38. George Arnold, *The Magician's Own Book or the Whole Art of Conjuring Being a Complete Hand-Book of Parlor Magic* (New York: Dick & Fitzgerald, 1857), 173.

39. Experiments to demonstrate binocular disparity included placing an object—your hand or a book—close to the face, closing first one eye and then the other, and noting the dissimilar images that resulted, with the left eye seeing more of the left-hand side of the object, and the right eye seeing more of the right-hand side. To understand the influence of binocularity on the perception of distance, sources recommended trying to snuff a candle or thread a needle with only one eye opened. See *Fireside Philosophy or Familiar Talks About Common Things* (New York: W. A. Townsend & Co., 1861), 298–9; "Wonders of the Stereoscope," *Eclectic Magazine,* July 1857, 377.

40. Burgess, *The Photograph Manual,* 197.

41. In some cases this relationship was reversed when the eye was called a natural camera obscura. *The Magician's Own Book,* 162. For an elaboration of the relationship between the camera lens and the lens of the eye, see Thomas Hankins and Robert Silverman, "The Giant Eyes of Science: The Stereoscope and Photographic Depiction in the Nineteenth Century," in *Instruments and the Imagination* (Princeton: Princeton University Press, 1995), 148–177.

42. See "The Stereoscope," *Godey's,* October 1852, 341. "The structure of the eye has at all times been quoted as one of the most beautiful illustrations of design and natural mechanism, and certainly the additional discoveries which we may expect to be disclosed by the Stereoscope will not diminish our wonders at the minute and beautiful arrangements by which external pictures are painted on the mirror of the mind."

43. George Arnold, *The Magician's Own Book,* 174–175.

44. "The Stereoscope, Pseudoscope, and Solid Daguerreotypes," *Photographic Art-Journal* 3, no. 3 (1852): 175.

45. Susan Sontag, among others, describes this relationship in her essay "The Image-World." Susan Sontag, *On Photography* (New York: Farrar, Straus and Giroux, 1977), 154. For another reading, see Williams, "Corporealized Observers," 1995.

46. Robert Hunt, "The Stereoscope," *Art Journal (London),* March 1856, 118. Similarly, *Eclectic Magazine* praised the stereoscope as "the means of realizing beneath our eyes all the kingdoms of

the world and the glory of them . . . not as in any picture, but standing out in all its solidity and reality, as if we were looking out of a window." "Stereoscopic Journeys," *Eclectic Magazine,* April 1857, 560.

47. Henderson, "Photography and the Stereoscope," 162.

48. *The Stereoscopic Magazine,* vol. 1. Earlier that year Reeve had published *Teneriffe, an Astronomer's Experiment; or, Specialities of a Residence above the Clouds,* the first book to be illustrated with stereographs. For information on Reeve and his commitment to stereographic illustration, see Amy E. Stark, "Lovell Augustus Reeve (1814–1865), publisher and Patron of the Stereograph," *History of Photography* 5, no. 1 (1981).

49. *The Stereoscopic Magazine* 6, 7.

50. Bierstadt, *Stereoscopic Views among the Hills of New Hampshire* (New Bedford, Mass.: 1862), n.p.

51. "The Progress of Photography," *Art Journal (London),* August 1, 1853, 181.

52. "Stereoscopes and Photographs," *Home Journal,* October 1 1859, 2.

53. "Photography as Applied to Business," *Photographic Times* 1, no. 4 (1871): 49.

54. G.A.S., "Origin of the Stereoscope," *Notes and Queries* 2 (fourth series) (1868): 465.

55. *Anthony's Photographic Bulletin,* November 1, 1870, 208.

56. Wade and Swanston, *Visual Perception: An Introduction,* 96.

57. The framework of these concluding remarks is indebted to Timothy Lenoir's Introduction to *Inscribing Science: Scientific Texts and the Materiality of Communication* (Stanford: Stanford University Press, 1998) and to Geoffrey Winthrop-Young and Michael Wutz's introduction to Kittler, *Gramophone, Film, Typewriter.*

58. "Iconography and Intellectual History: The Halftone Effect," in Neil Harris, *Cultural Excursions: Marketing Appetites and Cultural Tastes in Modern America* (Chicago: University of Chicago Press, 1990), 307.

6 Sinful Network or Divine Service: Competing Meanings of the Telephone in Amish Country

Diane Zimmerman Umble

The coming of the telephone to Lancaster County, Pennsylvania, captured popular attention and prompted local journalists to marvel at the power of this new medium. The county was home to enthusiastic promoters who founded and developed fourteen Bell-affiliated and independent telephone companies from 1898 to 1912. Those boosting telephone service appealed to farmers particularly: "The telephone is one of the most profitable business agencies that a farmer can employ," wrote one journalist in a small town newspaper. In addition to keeping farmers "in constant communication with the markets," telephony "provides a sitting room for the community where families can assemble and discuss the events of the day without the inconvenience of travel or loss of time, and in sickness and emergencies, it renders a divine service."

This metaphor of divine service likely troubled Old Order Mennonite and Amish residents of the county. As farmers, they appreciated the potential benefits of telephone access, but they took measure of the telephone from a unique perspective. An anti-telephone tract that circulated throughout the region urged that while "the telephone seems to be handy in many ways for people to know everything quickly," such knowledge was a worldly thing, making the telephone "a sinful network." Despite the promises trumpeted by telephone promoters, Old Order Mennonite and Amish people had grave misgivings about the impact of telephony. One Mennonite man recalls an elderly father scolding, "There goes the devil's wires."[1]

Divine service or sinful network? The contrasting characterizations of the telephone suggest that the meaning of telephony was disputed in the early years of the twentieth century within the particular contexts of Lancaster County. Accounts of new media often fail to acknowledge fully such disputed meanings. As Carolyn Marvin argues, the study of new media should approach media not as fixed objects with homogeneous effects, but as "constructed complexes of habits, beliefs, and procedures embedded in elaborate cultural codes of communication." The different responses of particular social groups help to reveal "issues crucial to the conduct of social life; among them, who is inside and outside, who may speak, who may not, and who has authority and may be believed."[2] Marvin's vocabulary of inside and outside, of authority and belief holds particular resonance when considering religious communities. This chapter describes the "telephone troubles" among the Old Order Mennonites and Amish. These painful, divisive debates about "who may speak [on the phone and] who may not" rumbled through the communities and severed some members from their churches. The "telephone troubles" provide a window into the dynamic interactions between new media and culture amid social and technological change. In particular, Old Order resistance to telephony, which emerged in the course of the troubles, demonstrates the role of new media in the ongoing formation of identities. On one level, telephonic communication put the religious community in jeopardy by remaking the practices of its separation from the larger world of which it was part. Old Order Mennonites and Old Order Amish certainly saw it that way when they decried a sinful network. On another level, however, the question regarding telephones formed only one part of an ongoing interrogation of the character and role of communication generally in the life of the religious community. The telephone helped mark but did not itself alone foment questions regarding the privileging of domestic, local, and oral communications, the habits of correspondence among far flung groups of believers and their elders or leaders, or the much debated practice of *ex*communication as a form of social control.

If telephony in rural American was uniquely shaped by "the socioeconomic geography of agriculture, the strong cooperative tradition in many regions, the long-standing customs of visiting and sharing labor," and "the gendered division of labor on the farm," in Lancaster County these forces were themselves shaped by religious contexts.[3] Mennonite and Amish people, whether they used, adapted, or fully resisted the new medium, acted in keeping with the identity of their community as a sort of switchboard, a crossing point, for identities that were at once deeply religious, necessarily local, and self-consciously rural and agrarian.

The Old Order Repertoire

By the 1880s, the Mennonites and Amish were distinguished by their German dialect, plain dress, and nonconformist practices and were respected as skillful and hard working farmers. During the third quarter of the nineteenth century, the Amish across the United States gradually divided into two groups, the Amish Mennonites or "Church" Amish (progressives who built meeting houses for worship) and the Old Order Amish or "House Amish" (traditionalists). Lancaster County's Amish settlement, the oldest in North America, was a stronghold of traditionalism. The Amish maintained their use of the German language, social exclusion of excommunicated members (shunning), use of horse and buggy for transportation, and worship in homes.[4]

The Mennonites also differed across a traditional and progressive continuum. In 1893, Bishop Jonas Martin led traditionalists out of the main body of the Mennonite church to form the Old Order Mennonites. To an outsider in 1880, Amish and Mennonite plain garb made them appear indistinguishable.[5] But Mennonite adoption of Sunday schools and participation in mission activity, along with the adoption of the English language, reflected a fundamental shift in worldview to those who claimed a hold on tradition. My focus here is on two German-speaking, horse and buggy, Old Order groups who suffered divisions over the telephone and who thrive today: the Old Order Amish and the Old Order (Groffdale Conference) Mennonites.

Scholars analyzing the development of Old Order movements within the Amish and Mennonite communities at the dawn of the twentieth century argue that the notion of *Gelassenheit* captures the essence of Old Order social repertoire.[6] *Gelassenheit* is a term used by early Anabaptists to communicate the ideal of yielding completely and unselfishly to the will of God. It means submission—yielding to higher authority—to God, the church community, elders, ministers, parents, and the tradition. In practice, *Gelassenheit* demands obedience, humility, submission, thrift, and simplicity. One "gives up" or "gives in" in deference to another for the sake of community. *Gelassenheit* is the standard for social relationships both within and beyond the group.

Faith, in the spirit of *Gelassenheit,* is not expressed in words. The church is not a set of doctrines. Rather, in the Old Order view, faith is expressed in the "way of life." The community is the expression of faith in the world, and eternal life is attained through the maintenance of the redemptive community—small, rooted in the land, mindful of its traditions, nonconformist, and separate from the world. Membership in the church entails submission to the *Ordnung,* or code of conduct. Prior to joining the church, new members are instructed in the content of the Confession of Faith and the *Ordnung.* When they join, they promise to submit to the congregation and its *Ordnung* for the rest of their lives. They are reminded of the cost of breaking their vow: They can be excommunicated and shunned. The *Ordnung* functions as a means of regulating change while maintaining essential community values. When the congregation faces new issues, the leaders and members discuss it. As consensus develops, the position is "grafted into the *Ordnung.*"[7]

At the turn of the twentieth century as well as today, communication among the Older Order Amish and Mennonites was infused with a spirit of *Gelassenheit* and practiced through community rituals of worship, silence, work, and visiting that were anchored in the home.[8] Patterns of communication built and maintained strong, primary relationships within the circle of church life. Even when Old Order people interacted with their "English" neighbors, their dress, dialect, and church membership reminded them where they belonged. With the arrival of telephones, new ways of communicating threatened to change the face-to-face character of communication and orient communication away from the home toward the outside world.

Telephone Service in Lancaster County

Although Bell Telephone service was established in Lancaster in 1879, it was not until the expiration of key Bell patents in 1893 that competing companies were founded in Lancaster and the surrounding rural areas. By 1898, the Independent Telephone Company competed head to head for city subscribers and wooed rural subscribers who were already clamoring for service, but were underserved by Bell.

Rural areas were also served by a variety of smaller organizations both formal and informal. Farmers in some areas organized their own private lines, stringing wire from fence post to fence post, linking four to six neighbors on a single party line. In 1898, a village newspaper reported: "In many parts of the country farmers have established among themselves a telephone system covering eight or ten miles of wire, the wire used being barbed wire fences. The middle wire of the fence is used, and the farmers are able to converse with each other without difficulty, thus relieving a part of the lonesomeness which forms a chief objection to farm life."[9] Hundreds of farmers' lines were organized throughout the countryside in the early years. In some cases, groups of local farmers and businesses organized to build a local line and later petitioned for connection with a larger company. In other cases, local parties formally organized and chartered a company to provide local and long distance service.

By 1899, New Holland and the surrounding communities had access to Independent lines as well as Bell lines. During 1899, the Independent Company ran lines from Lancaster to Ephrata, Intercourse, Gap, Hinkletown, Blue Ball, East Earl, and Weaverland—towns in the heart of the Old Order Mennonite settlement. By 1900, Independent lines also extended into the Amish settlement in the southern part of the county.[10] In response to vigorous independent competition in rural markets, Bell liberalized its interconnection policies. By 1910, more independent telephone companies were interconnected with Bell than remained outside the system.[11] By 1912, ten different telephone companies provided service to Lancaster county. Competition for local subscribers was intense as companies vied for loyal customers through coverage and advertisements in the pages of local newspapers.

Lancaster County experienced rapid and vigorous telephone development in the years between 1900 and 1912. These local telephone companies were organized, not by outsiders, but by the leading farmers and businessmen in local towns and villages, including Amish Mennonites, Mennonites, and at least one Old Order Mennonite. In 1902, a majority of the organizers of the Conestoga Telephone and Telegraph Company to the east of New Holland were members of the Conestoga Amish Mennonite congregation. The Enterprise Telephone and Telegraph Company, also organized in 1902, included several Mennonites and one Old Order Mennonite on its board. The Intercourse Telephone and Telegraphy Company, founded in 1909 to compete with Enterprise, also included Mennonites on its board.[12] When the telephone debates came to a head within the Old Order Mennonite community between 1905 and 1907 and within the Old Order Amish community in 1909 and 1910, Old Order people had access to both Bell and Independent services for local and long distance service.

Divine Service

To proponents in Lancaster County at the turn of the century, telephone ownership served as a mark of the progressive farmer or the efficient rural businessman, doctor, or lawyer. The pages of one weekly village newspaper, the New Holland *Clarion,* hailed the telephone for providing efficient access to current information: market reports, weather reports, and transportation schedules. The telephone facilitated doing business by preventing unnecessary trips to town and handling emergencies quickly. Editors promoted telephone service for its potential contributions to the growth, profits, and efficiency of local businesses.

Telephone company advertising in the village weekly newspapers amplified these themes by emphasizing the value of the telephone in times of emergency: accidents, fires, illness, stolen horses, mad dogs, robbers, and threatening weather. One Enterprise Telephone Company advertisement provided nine reasons that readers, both men and women, need a telephone. "Some of the Reasons You Need a Telephone" begins with "So your wife can use it daily, to order her meat and groceries. You can get at once into communication with your home when you are away. . . . If every clock in the house stops, you can get correct time from central." The list includes summoning help in cases of illness, fires, or accidents, facilitating social arrangements, and obtaining market prices. The copy also claims, "You can increase your circle of desirable acquaintances as a telephone in the home gives you a social distinction in the country."[13] Indeed, one writer claimed that the telephone could actually lengthen life: "By use of the telephone, more work can be crowded into one day, . . . increasing the length of one's life, as after all, what really counts is what we actually accomplish."[14]

As telephone promoters realized that women were an important market, illustrated advertisements began showing women enjoying the "wonderful comfort and pleasure" of rural connections. "I just called up to ask you over this evening," the copy reads. Not only did the telephone help this farm family save money, it also makes it possible to be in contact with friends "only ten seconds away" (figure 6.1). Another advertisement models visiting by telephone. The copy reads, "How pleasant it is to make a telephone visit to relatives or friends. The distance only adds enchantment to your chat."[15] Not all telephone socializing, however, was deemed appropriate. During the summer of 1906, a series of articles and letters in the *Clarion* vented the frustration many felt about those who listened in on party lines, participated in spooning on the line, or used abusive, profane, or obscene language. To curb bad behavior, local directories often printed rules and instructions for the proper use of the telephone.[16]

For its proponents, the telephone was associated with profit, comfort, and pleasure. It widened the world for rural people, providing potential connections to centers of power, information, and culture. The telephone was an instrument of pleasure and progress, a mark of success, and even, on occasion, a medium of divine service. One telephone advertisement sums up its meaning: "The old order of things has passed. To be modern is to have a Bell Telephone. To have a telephone is to *live*" (figure 6.2).

Old Order people were not blind to the practical benefits of the telephone, but they were deeply suspicious of its social and spiritual implications. What really counted for them was not defined in terms of the modern world or personal accomplishment. And to characterize the telephone as divine was unthinkable.

The Sinful Network: The Old Order Mennonite Debates

At the turn of the twentieth century, Bishop Jonas Martin was the acknowledged leader of the Old Order Mennonites in Pennsylvania and had wide influence among Old Order

Figure 6.1   "I Called up to Ask You over This Evening." Reprinted from the New Holland *Clarion* 3 (August 1912): 2. From the collection of the Lancaster County Historical Society, Lancaster, Pennsylvania.

Figure 6.2   "Getting over the old stile." Reprinted from the New Holland *Clarion*, 26 May 1912, 6. From the collection of the Lancaster County Historical Society, Lancaster, Pennsylvania.

Mennonite groups across North America. His correspondence with other conservative bishops and ministers in Ontario, Ohio, Indiana, Virginia, and Pennsylvania reveals the heated and sometimes personal terms of their debates. An antitelephone tract circulated among them at the turn of the century characterizes the telephone as a "sinful network" that leads to association with "unbelievers." The anonymous writer claims that telephone conversations amount to "vain jangling" or "useless talking" that detract from "spiritual" activities. Finally, the writer exhorts, "Do you trust this network to save a life? . . . Look to Jesus for all you need."[17]

Old Order rules of separation limited members' associations with persons from outside the church, lest they be tempted to behave as worldly people behaved and to adopt worldly values. Rules preventing business associations beyond the church protected members from the tensions of divided loyalties and from participation in the legal system. Amish and Mennonite church disciplines of the day also prohibited members from swearing oaths, but State annual reports on telephone company activity required company directors to "swear" to the truth of these reports by their signatures. This presented an ethical problem for church members. Directors of the Enterprise Company crossed out the word "swear" and wrote in the word "affirm" to address the conflict between church rules and business expectations. Many Mennonite and Amish congregations discouraged or prohibited stock ownership.

Not only were business associations potentially tempting, but the private nature of communication by telephone appeared to hold temptations as well. "Spooning" and "listening in" were socially recognized practices associated with telephone use, particularly among young people and women. Neither practice conformed to Old Order expectations of humble and submissive speaking and living.

But perhaps the greatest temptation of the telephone was its challenge to faith in God. To place trust in a machine for protection, as the telephone proponents advocated, seriously undermined recognition that faith in God alone was sufficient. In Ontario, Ohio, Indiana, and Iowa, ministers were divided over whether to permit their members to own telephones and telephone company stock. In the years 1905 to 1909, some Old Order Mennonites had telephones in their homes and owned stock in telephone companies. In Lancaster County, for example, Old Order member Noah Nolt owned stock in the Enterprise Telephone Company of New Holland and served on the board of directors in 1905. Nevertheless, Bishop Jonas Martin preached from the pulpit against telephone ownership. The letters from the ministers who sought his advice reflect the great tension generated by the issue of the telephone. In some congregations in the Midwest, members who had telephones were prohibited from taking communion, an action just short of excommunication. In other congregations, members were expected to remove their telephones before they could participate in communion.

In March 1905, Bishop Martin received a letter from Lancaster County natives who had moved to Iowa. In addition to descriptions of health, crops, and weather, the writer mentions difficulties in the local congregation over the telephone. He reports that the congregation had not held communion for two years because of dissension over the telephone.[18] Ministers in Iowa were against telephone ownership, but some of their members were in favor of the new technology.

Although Bishop Martin preached against telephones, the Pennsylvania churches had not ruled them out completely. Martin did take a stand against the ownership of telephone stock, but his position was confusing. Those on both sides of the issue pressed Martin for clarification of his position. In August 1905, Martin wrote to Ohio leaders that he was "very sorry that the things are so badly understood." He reports, "I am a witness against them. But we have not cut them off." Martin explains that he hopes that by exercising love, rather than drastic measures of separation, members could be convinced to give up their telephones. Thus he did not prevent telephone owners from joining the church or taking part in communion in his district.[19]

The bishop's extensive correspondence illustrates the dynamics of the debates taking place. Both sides appealed to the faith of the fathers in support of their arguments. Proponents argued that maintenance of peace and forbearance were the highest goals. Furthermore, they resented younger leaders who ignored the counsel of their elders in the church. Those opposed to the telephone felt just as strongly that they were protecting the tradition of the fathers by rejecting an "instrument of strife" that brought pride and worldliness into the church. Advocates of toleration diagnosed the problem to be lack of commitment to love and peace within the community, while opponents of compromise saw the telephone itself as an instrument of discord. Old Order Mennonites were thus torn between peace and purity.

The Pennsylvania churches tried to find a middle ground. Bishop Jonas Martin continued to preach against telephone ownership, and his public desire was to make telephone ownership a test of membership. But unlike his Indiana cousin, Jonas Martin was unwilling to override the judgment of his elders in church leadership. At the spring ministers' conference in April 1907, Martin finally made a formal concession to telephone ownership after the eldest and most conservative deacon in the district pleaded with him to avoid a church split. Deacon Daniel Burkholder himself opposed telephone ownership and agreed with Jonas Martin's position. But Burkholder argued that Martin had led one division (over Sunday schools) and that was enough for one lifetime. Judging from the developments in Ohio and Indiana, Burkholder felt that taking a strong antitelephone position would split the Pennsylvania churches, and so the Pennsylvania conference crafted a compromise.[20]

The ruling announced on April 5, 1907, tried to satisfy both the progressives and the conservatives by establishing specific conditions for telephone use: 1) church members could not enter hotels to use telephones, 2) church leaders could not own telephones, and 3) those nominated for leadership positions in the future could not own telephones. The agreement stated that while the leaders did not support telephones and felt members were better off without them, members who owned telephones could participate in communion and would not be excommunicated. In effect, the ruling limited the pool of potential leaders to those who were most conservative, or at least those who were willing to give up the telephone if called to ministry. Church leaders hoped that the telephone could be "preached away." In order to keep all members connected to the community, membership had to be defined to include those who had fallen to the temptation of telephones.

The Pennsylvania Ruling, as it is called, proved confusing to many. Jonas Martin's antitelephone attitude was well known, and those who expected him to support a hard line were disappointed. Others took comfort in the compromise. By submitting to his elders, against his personal feelings, Jonas Martin had averted a major division in his own district. One local minister, Jacob B. Weaver, wanted to exclude completely the telephone, believing that members should not be allowed to "partake in such a worldly thing."[21] Weaver was so distraught over the compromise that he never preached or attended church services again. On the other hand, an Old Order Mennonite called "Gentleman" Harry Martin refused to sell his telephone stock. As a result, he was expelled and later joined a nearby Lancaster Conference Mennonite church.

The Ohio and Indiana groups suffered a major division over the telephone a month later on May 17, 1907. John Martin, along with ten ministers and four deacons, walked out of the spring meeting of the Ohio and Indiana Conference in 1907. While the majority of the leadership followed John Martin out of the conference, two-thirds of the Indiana laity followed those elders in favor of toleration.[22]

Some Old Order Mennonites were willing to risk church division over their understanding of the telephone. They worried over the influences the telephone might bring into the home and the church. Accommodating those who wanted the telephone out of personal weakness was tantamount to leading them into temptation. The purity of the church and the home was at stake. Strong desires for purity clashed with desire for "bearing one another in love." John W. Martin defied the elders in Indiana for the sake of purity. His cousin, Jonas Martin, submitted to Pennsylvania elders for the sake of peace.

The Pennsylvania Ruling was observed by the Groffdale Conference of the Old Order Mennonites for eighty-seven years. In 1994, after three years of debate, Old Order Mennonite leaders declined to enforce the telephone prohibition for clergy in the 1907 Ruling. A small group who objected to the change left the conference and formed new congregations. While the main body of Old Order Mennonites no longer observes the Pennsylvania Ruling, smaller groups in Pennsylvania and Missouri remain committed to the 1907 guidelines today.

"Putting Away" the Telephone: The Old Order Amish Troubles

Among the Old Order Amish, the telephone played a similar role to the one it played among the Old Order Mennonites. The Old Order Amish believe that the telephone was a principal issue behind the 1910 division of the Amish church that resulted in the loss of one fifth of its membership.[23] Prior to 1910, Amish leaders had taken no firm position on the telephone, and some Amish families had telephones in their homes. According to oral sources, these farmers were linked to families in the immediate neighborhood by party lines, and possibly were not connected to organized telephone companies.

The Amish kept few written records of church matters, since information was passed down from generation-to-generation through face-to-face encounters. Thus, the story of the telephone debates comes primarily from the Amish oral tradition. Two themes are present in the oral accounts. One theme describes differences in willingness of Amish men to "put away" the telephone. Some church members were willing to give up their telephones in submission to the decision of the church leaders; others were not. The other theme features an account of difficulties that arose when women gossiping on a party line were overheard by the object of their gossip who was listening in on the conversation. "This made quite a stink," writes an Amish man; "then the Bishops and the ministers made out if that is the way they are going to be used we would better not have them."[24]

Whether the catalyst was the gossip of the women or the stubbornness of the men, the dissension within the community had to be addressed before communion could be held. The telephone became one of the issues for examination at the fall 1909 ministers' conference prior to communion. When local ministers could not come to consensus on several problems on the agenda, ministers from Mifflin county were brought in to moderate. In addition to the telephone question, oral sources suggest that disagreements about shunning and meeting houses were also on the agenda.

After joint consultations, the ministers announced their intentions to hold to the "old order" followed by their forefathers. Their statement at the fall 1909 meeting reiterated support for the practice of the excommunication and shunning of those who were unfaithful to church rules. Furthermore, they made it clear that the telephone had no place in the Amish home. An Old Order Amish historian writes that the 1910 split was caused by "indifferent views in church discipline, most concerning newly invented contraptions that our conservative church leaders could not tolerate."[25]

Through Amish eyes, the telephone in the home manifested traits that were contrary to the spirit of *Gelassenheit*. Amish explanations of the telephone ban suggest that the telephone was not a necessity. The telephone was of "the world" and led to association with outsiders. Personal ownership of the telephone contributed to individualism and pride rather than humility. Furthermore, women were tempted to use it for gossip, which disrupted social harmony. In general, the use of the telephone did not conform to the time-honored principles of nonconformity and separation from the world. By their decision, Amish leaders reaffirmed and reapplied the tradition of the past to a challenge of the present. In the course of affirming the "old order," they excluded the telephone from the home. At the same time, they reiterated the costs of unfaithfulness—excommunication resulting in shunning.[26]

Historians of the departing group, now called the Beachy Amish, attribute the split to its unwillingness to adhere to the strict application of shunning, in particular, the obligation to shun someone who left the Amish church to join a nearby Mennonite congregation. In the fall of 1909, thirty-five families, representing one fifth of the membership of the Lancaster settlement, announced their intentions to withdraw if their more moderate views on shunning was not respected. According to an Amish publication, the dissenters' position was communicated in the form of an unequivocal "demand." The new group began holding its own services in 1910.[27] In the years to come, the Beachy group adopted the telephone, the tractor, electricity, and automobiles. They also began meeting in church buildings instead of homes. All these practices were prohibited by the Old Order Amish *Ordnung*. Today, both sides remember the split as a painful separation.

Given the contrasting perspectives, Old Order Amish historians maintain that the telephone was the "handle" on which the division turned.[28] Whether the issue was the telephone or differences over shunning, the dissenters stood as an example of unwilling-ness to submit to the "ordering" of the tradition and of the elders. The Amish argue that the dissenters "demanded" respect for their views on the basis of threats to withdraw. To the Amish, the Beachy group represents what happens when a "worldly" spirit creeps in, and the telephone was implicated in that "worldliness."

What is not included in the accounts of either side is the fact of the extension of tele-phone service into their midst at the time of the "troubles." In the summer of 1909, sev-eral progressive, non-Amish citizens of Intercourse, the village in the heart of the Amish community, organized a telephone company with considerable fanfare in the local press. Several months after it organized, the company entered into agreements with the Bell sys-tem to connect with Bell trunk lines, thereby affording connections to Lancaster city and points across the county. By the fall of 1909, it was possible to connect local family lines to the "outside world." That same fall, the telephone issue came before church leaders for debate. The telephone was at their door, and it was no longer just local.

The Amish approach to technology has been one of selective adoption, determined on the basis of whether or not a particular innovation preserves or threatens community life. In the 1890s, for example, the grain binder was readily accepted and the Amish were among the first to own the new machine.[29] Rules governing use of the telephone set precedents that made rejecting the use of electricity and technologies such as radio and television much easier. Telephone use itself has not been banned, but rather the installa-tion and private ownership of a telephone in the home. In the intervening years, the Old Order Amish have modified the rules about telephone ownership in the face of economic change. Today telephones are found in or near many Amish businesses, although the ban on telephones in the home persists.[30]

The telephone was perceived as a threat because it entered the home at the heart of Amish faith and life, in essence, sacred space. The telephone stood as both a symbolic and a physical connection to the outside world, and it opened the home to outside influence and intrusion. This new technology removed communication from a face-to-face con-text, where the often silent discourse of community occurred, and replaced it with the potential for individual links with sources of information from outside. These new links were unmediated or filtered by the style, rhythms, and rituals of community life. Tele-phone use decontextualized communication and, thus the Old Order communicator.

The spirit of the telephone debates also posed a challenge to Old Order authority structures. Oral accounts on both sides of the debates describe "unwillingness to submit" to church leaders as part of the problem. This challenge to *Gelassenheit* could not be over-looked. For the Old Order Amish, the linking of the two issues—the telephone ban and social avoidance of wayward members (shunning)—stands as a decisive corrective for any who were tempted to set their sights too far beyond the context of community. The order of community had to be preserved, even at the cost of losing some members. The telephone debates show how the repertoire of a religious community can be tested by the introduction of a new medium. Old Order struggles demonstrate how culture shapes the meaning of new technology within the dynamic contexts of cultural self-definition. And the resistance of the Old Order Mennonites and Amish communities illustrates the efforts two groups have made to adapt a range of communication practices to the service of community values.

Notes

1. I reviewed the *Ephrata Review* (ER) from 1900 to 1914 and the *New Holland Clarion* (NHC) from 1883 to 1914 for coverage about the telephone. The quote is from ER 6 (November 1914): 9. The antitelephone tract was published along with other papers and letters of Old Order Mennonite bishop, Jonas Martin (1839–1925), by Amos B. Hoover, ed., *The Jonas Martin Era* (Denver, Penn.: Muddy Creek Farm Library, 1982), 811–812. Warren Weiler recounted the story about the "devil's wires" in a personal interview. Weiler's ancestors founded the Enterprise Telephone and Telegraphy Company in 1902. Old Order informants are not identified at their requests.

2. Carolyn Marvin, *When Old Technologies Were New: Thinking about Electric Communication in the Late Nineteenth Century* (New York: Oxford Press, 1988), 8, 4. In contrast to those who approach communication as transmission, I take a cultural approach, defining communication as a process through which humans symbolically construct meaningful worlds to live in. James Carey, *Communication as Culture: Essays on Media and Society* (Boston: Unwin Hyman, 1989) advocates a cultural approach to communication "directed not at the extension of messages in space but toward the maintenance of society in time, not the act of imparting information but the representation of shared beliefs" (p. 18). Claude S. Fischer's *America Calling: A Social History of the Telephone to 1940* (Berkeley: University of California Press, 1992) is a comprehensive sociological analysis of the coming of the telephone and the automobile, and includes an extensive bibliography of telephone research. See Lana Rakow, *Gender on the Line: Women, the Telephone, and Community Life* (Urbana: University of Illinois Press, 1992) for analysis of gender and the telephone.

3. Ronald R. Kline, *Consumers in the Country: Technology and Social Change in Rural America* (Baltimore: Johns Hopkins University Press, 2000), 40–41. See especially chapter 4 "(Re)Inventing the Telephone." Kline mentions resistance to the telephone ("relatively short lived and passive" [40]).

4. Cornelius J. Dyck, *An Introduction to Mennonite History,* 2d ed.(Scottdale, Penn.: Herald Press, 1981) provides historical background on the Anabaptist movement. Calvin Redekop, *Mennonite Society* (Baltimore: Johns Hopkins University Press, 1989) writes on Mennonite society, past and present. The classic anthropological study of the Amish is John A. Hostetler, *Amish Society,* 4th ed. (Baltimore: Johns Hopkins University Press, 1993) and includes recent scholarship on Anabaptist origins. Donald B. Kraybill, *The Riddle of Amish Culture* (Baltimore: Johns Hopkins University Press, 1989) provides sociological analysis of the Amish. Steven M. Nolt, *A History of the Amish* (Intercourse, Penn.: Good Books, 1992) provides an historical account of North American Amish. The Old Order Mennonites and the Amish share the same Anabaptist origins. For an account of the Old Order movements, see Theron F. Schlabach, *Peace, Faith, Nation: Mennonites and Amish in Nineteenth-Century America* (Scottdale, Penn.: Herald Press, 1988), 201–229.

5. Donald B. Kraybill, "At the Crossroads of Modernity: Amish, Mennonites and Brethren in Lancaster County in 1880," *Pennsylvania Mennonite Heritage,* 10 (January 1987): 2–12.

6. Sandra Cronk, "*Gelassenheit:* The Rites of the Redemptive Process in Old Order Amish and Old Order Mennonite Communities," *Mennonite Quarterly Review* 55 (1981): 5–44. Kraybill (1989) and James C. Juhnke, *Vision, Doctrine, War—Mennonite Identity and Organization in America 1890–1930* (Scottdale, Penn.: Herald Press, 1989) acknowledge Cronk's influence on their analyses. Kraybill reports that Amish themselves find her analysis valid (268). Cronk paints the North American context out of which the Old Order Division grew. She argues that three North American phenomena introduced an alternate "redemptive process": 1) growing technological and industrial development, 2) the political ideology of democracy, and 3) pietism and Evangelical Protestantism. Cronk's work shapes my analysis as well. Steven D. Reschly, "Alternative Dreams and Visions: The Amish Repertoire of Community on the Iowa Prairie, 1840–1910," Ph.D.diss., University of Iowa, 1994, 7, coins the phrase "repertoire of community" consisting of "channels of consciousness and habits" that guide and limit possible ranges of attitudes and actions available to members of a community. He argues that a community fashions itself in dynamic relationship with conceptions of the "world" beyond the community.

7. Kraybill, *The Riddle,* 96.

8. For a more complete description of how rituals of worship, silence, work, and visiting order community life, see Diane Zimmerman Umble, *Holding the Line: The Telephone in Old Order Mennonite and Amish Life* (Baltimore: Johns Hopkins University Press, 1996), chapter 3.

9. NHC 11 April 1898, 3.

10. The development of telephone service and the growth of independents in Lancaster County parallels nationwide telephone trends. Nationally, as in Lancaster County, Bell gave primary attention to urban development until forced to confront strenuous independent competition in rural markets. Lancaster County data was collected from A.T.&T. Exchange Statistics 1880–1907 (Warren, N.J.: A.T.& T. Archives): Box 7 and "Reports of Telegraph and Telephone Com-

panies to the Auditor General and the Department of Internal Affairs 1881–1916" (Harrisburg: Pennsylvania State Archives), RG.1.

11. Robert Garnet, *The Telephone Enterprise* (Baltimore: Johns Hopkins University Press, 1985), 131.

12. Brief histories of the companies founded in Old Order regions, including the Conestoga Telephone and Telegraph Company, the Enterprise Telephone and Telephone Company and the Intercourse Telephone and Telegraph Company, are covered in chapters 5 and 6 in Umble.

13. NHC 19 August 1911, 5.

14. ER 23 August 1912, 4.

15. NHC 11 May 1912, 5.

16. Umble, *Holding the Line,* 83–84.

17. Hoover, *Jonas Martin Era* 811–812.

18. Ibid., 164.

19. Ibid., 675–676.

20. Ibid., 687.

21. Ibid., 509.

22. Ibid., 897–898.

23. Kraybill, *The Riddle,* 144.

24. John K. Lapp, "Remarks of By-Gone Days: A Few Remarks of Old Times." Pamphlet (Gordonville, Penn.: Gordonville Print Shop, 1986): 7.

25. Joseph F. Beiler, foreword to *Amish and Amish-Mennonite Genealogies,* compiled by H. F. Gingerich and R. W. Kreider (Gordonville, Penn.: Pequea Publishers, 1986), xiv.

26. The Lancaster Amish settlement was not alone in its struggles over the telephone. Desire to participate in Iowa telephone companies among Amish church members precipitated long-running debates in the Upper and Lower Deer Creek congregations. See Roy A. Atwood, "Telephony and its Cultural Meanings in Southeastern Iowa, 1900–1917," Ph.D. diss., University of Iowa, 1984: 362–367. In 1906, an attempt to diffuse some of the resistance resulted in an Amish petition to the local mutual telephone company. They asked to be allowed to subscribe to telephone service as "donors" instead of becoming stock-owning "members of the company." The company granted their request, and Amish subscribers could "donate" rather than buy shares in the local company in exchange for service. Nevertheless, the issue went unresolved in both Iowa congregations for several years.

27. Oral sources on the Beachy side of the debate suggest that even the splinter groups may have prohibited telephones in the home. In two cases, descendants of those who left the Old

Order recall that their families asked the bishop for permission to have telephones in 1916 and in 1923. Those who tell the Beachy story argue that few people had telephones in those days. See Aaron S. Glick, "Pequea Amish Mennonite Church Twenty-fifth Anniversary." Mimeo, 1987. See also Elmer S. Yoder, *The Beachy Amish Mennonite Fellowship Churches* (Hartville, Ohio: Diakonia Ministries, 1987).

28. Kraybill, *The Riddle,* 143.

29. Gideon L. Fisher, *Farm Life and Its Changes* (Gordonville, Penn.: Pequea Publishers, 1978).

30. Cellular telephones pose a new challenge for Amish leaders. Because cell phones are easily concealed, it is difficult to know just how many are in use among the Amish.

7 Souvenir Foils: On the Status of Print at the Origin of Recorded Sound

Lisa Gitelman

When Thomas Edison began demonstrating the phonograph to eager audiences in 1878, he promoted the machine—and the American public received it—as an invention that would revolutionize print. The "speaking phonograph" or "talking machine," as it was first known, was able to record as well as to reproduce sound. (It would not be widely adopted as an amusement device for musical playback until the mid-1890s.) Throughout 1878 and for more than the decade to follow, phonographs seemed to offer an unprecedented, excitingly modern connection between aural experience and inscribed evidence, between talk and some new form of text. The recording surface was originally tinfoil. Air set in motion by the production of sound acted upon a diaphragm connected to a stylus; the stylus indented the foil to "capture" what the inventor Thomas Edison called "sounds hitherto fugitive" for later "reproduction at will."[1]

At demonstrations throughout the United States and abroad during 1878, audiences greeted the phonograph with both enthusiasm and skepticism. On the one hand, they marveled at the unprecedented phenomenon of recorded sound (a machine that speaks!). Many, however, felt disappointment when, after all the hyperbolic rhetoric surrounding the device, the early, imperfect phonograph produced only faint sounds obscured by scratchy surface noise. But however mixed their reactions may have been, audiences at the phonograph demonstrations regularly and eagerly took scraps of tinfoil home with them when the lectures were over. This chapter pursues those souvenir scraps of indented foil. Those primitive records were clearly meaningful to the women and men who sought them and who were probably asked at the breakfast table the next morning, "What does it say?" Without the phonograph for playback, the tinfoil records of course said nothing. Yet for the people who brought them home, the very same records clearly said *something*.

These sheets of foil were talismans of print culture. They were pure "supplement," in the language of literary study today, illegible and yet somehow textual, public and inscribed. Although themselves neither written nor printed, their apprehension became a necessary part of the social practices according to which printed and other text-objects were understood. They literally con*textualized*. Put another way, their status as curious new-media texts helped to inscribe the meanings of old-media textuality that had pertained and that would pertain in the future. Because of what they were—and more particularly because of what they were not—the production and collection of tinfoil records formed a new social experience of text and thereby of print. As such these scraps of tinfoil offer one way to inquire into the social meanings of printedness toward the end of the nineteenth century. For sure, these same meanings may also be glimpsed in other, attendant experiences of text and of print—in the ongoing construction of authorship, for instance, in the changing political economies of publishing, the additional subjectivities of late-century literacy, the shifting character and institutional status of textual criticism, and so on. The list is long. Against it tinfoil souvenirs can profitably be read as *foils* in the literary or the schoolbook sense. Historical characters important in part for what they were not, the foils defined by contrast, acting in mutual opposition to other characters—characters like authors, readers, publishers, and critics, but even more particularly the characters who "spoke" between quotation marks from the pages of American newspapers and in other publications.

Aided by the surrounding publicity, tinfoil records offered a profound and self-conscious experience of what "speaking" on paper might mean. Judging at least from contemporary quarrels between philologists and rhetoricians over the appropriate study of language or the concomitant popularity of "verbal criticism" and paeans to "good usage,"[2] tinfoil records were less causal agents of change than they were fully symptomatic of their time. The contemporary promotion of simplified spelling, the rampant competition between different shorthand systems, the widespread publishing of dialect and regional literatures, the scholarly worries about the "correct" pronunciation of Latin, the probable pronunciation of English by Chaucer and Shakespeare, and the appropriate "collection" of non-Western tongues, "authentic" black spirituals, folktales, and English ballads all reflect concern with the issue of "speaking" on paper. As Jon Cruz has shown in the case of collected spirituals, these concerns lie directly at the heart of "the modern hermeneutical orientation."[3] They are central to the disciplines and disciplinary practices of explaining culture, disciplines and practices that emerged with force toward the end of the nineteenth century, and whose emergence was aided by the structures of the American research university and attended by the birth of a canonical national tradition. This chapter tells the story of a few, fragile sheets of tinfoil because they offer another modest opportunity to peer into the concerns of their time. In particular, tinfoil souvenirs make visible certain anxieties regarding the medium of print. One undercurrent to the story that follows is the speculation (which surfaces in another form in chapter 8 below) that such anxieties can profitably be read against and within the disciplinary formations of the 1880s and 1890s.[4]

Thomas Edison was the first to demonstrate the phonograph in public, when he took his prototype to the New York City offices of the *Scientific American* magazine in 1877. There, witnesses reported, the phonograph greeted them and inquired after their health. They were fascinated by the apparent simplicity of the device; it was "a little affair of a few pieces of metal," not a complicated machine with "rubber larynx and lips." Wrapped around a cylinder rotated by hand, the tinfoil recording surface was impressed with indentations that formed "an exact record of the sound that produced them" and comprised what was termed "the writing of the machine." These words or "remarks" could then be "translated" or played back. Observers seemed for a time to believe that they themselves might translate, using a magnifying glass painstakingly to discern phonetic dots and dashes. But the really remarkable aspect of the device arose, one onlooker marveled, in "literally making it read itself." It was as if, "instead of perusing a book ourselves, we drop it into a machine, set the latter in motion, and behold! The voice of the author is heard repeating his own composition."[5] Edison and his appreciative audience clearly assumed that his invention would soon provide a better, more immediate means of stenography. Machinery, accurate and impartial, would objectively and materially realize the author's voice.

In statements to the press and later in his own article in the *North American Review,* Edison enumerated the use of phonographs for writing letters and taking dictation of many sorts, as well as for things like talking clocks, talking dolls, and recorded novels. Music was mentioned, but usually as a form of dictation: You could send love songs to a friend, sing your child a lullaby, and then, if it worked, save up the same rendition for bedtime tomorrow. In keeping with the important public uses of shorthand for court and legislative reports, the phonograph would also provide a cultural repository, a library for sounds. The British critic Matthew Arnold had only recently defined culture as "the best that has been thought and said in the world," and now the phonograph could save up the voices and sayings, Edison noted, of "our Washingtons, our Lincolns, our Gladstones." And there was plenty more to save. The American Philological Society, Edison reported to the *New York Times,* had requested a phonograph "to preserve the accents of the

Onondagas and Tuscaroras, who are dying out." According to the newspaper, only "one old man speaks the language fluently and correctly, and he is afraid that he will die."[6]

The contrast between *our* statesmen and the dwindling Onondaga hints that the phonograph was immediately an instrument of Anglo-American cultural hierarchy. It became party to habitual and manifold distinctions between an "us" and a "them." Drawing a similar distinction *Punch* magazine satirized in 1878 that the work of "our best poets" could be publicly disseminated by young women using phonographs, taking the place of the "hirsute Italian organ-grinders" who walked about the streets of London. The phonograph became at once instrumental to expressions of difference like these and suggestive of a strictly (because mechanically) nonhierarchical vox populi. One enthusiast proposed half seriously that a phonograph could be installed in the new Statue of Liberty, then under construction in New York Harbor, so it could make democratic announcements to passing ships. Reality seemed hardly less fanciful. With this remarkable device, published accounts made clear, women could read while sewing. Students could read in the dark. The blind could read. And the dead could speak.[7]

In January 1878 Edison signed contracts assigning the rights to exhibit the phonograph, while reserving for himself the right to exploit its primary dictation function at a later time. Exhibition rights went to a small group of investors, most of them involved already in the financial progress of Alexander Graham Bell's telephone. Together they formed the Edison Speaking Phonograph Company and hired James Redpath as their general manager. A former abolitionist, Redpath had done the most to transform the localized adult-education lecture series of the early American lyceums into more formal, national "circuits" administered by centralized speakers' bureaus. He had just sold his lyceum bureau, and he came to the phonograph company with a name for promoting "merit" rather than what his biographer later dismissed as "mere newspaper reputation."[8] The distinction was blurred, however, throughout the ensuing year of phonograph exhibitions. With so much distance separating the primitive phonograph's technological abilities from the ecstatic hyperbole that surrounded the invention in the press, phonograph exhibitors relied upon novelty in their appeals to audiences. Novelty, of course, would wear off, although it would take well into the summer of the following year to "[milk] the Exhibition cow pretty dry," as one of the company directors put it privately in a letter to Edison.[9]

The Edison Speaking Phonograph Company functioned by granting regional demonstration rights to exhibitors; individuals purchased the right to exhibit a phonograph within a protected territory. They were trained to use the machine, which required a certain knack, and agreed to pay the company 25 percent of their gross receipts. This was less of a lecture circuit, then, than a bureaucracy. For the most part, phonograph exhibitors worked locally; whatever sense they had of belonging to a national enterprise came from corporate coordination and a good deal of petty accountancy. Paper circulated around the country—correspondence, bank drafts, letters of receipt—but the men and their machines remained more local in their peregrinations, covered in the local press, supported (or not) by local audiences and institutions in their contractually specified state or area. The company set admission at twenty-five cents, although some exhibitors soon cut the price down to a dime. Ironically, no *phonographically* recorded version of a phonograph exhibition survives; the tinfoil records did not last long. Instead, the character of these demonstrations can be pieced together from accounts published in newspapers, letters mailed to Redpath and the company, and a variety of other sources, which include a burlesque of the exhibitions entitled *Prof. Black's Phunnygraph or Talking Machine*.

While the Edison Speaking Phonograph Company was getting on its feet, several of Edison's friends and associates held public exhibitions that paired demonstrations of the telephone with the phonograph and raised the expectations of company insiders. Charging theater managers $100 a night for this double bill, Edward Johnson toured upstate New York at the end of January. Not all of his performances recouped the hundred dollars, but in Elmira and Courtland, "It was a decided success," he claimed, and the climax of the evening was "always reached when the Phonograph first speaks." "Everybody talks Phonograph," Johnson reported, on "the day after the concert and all agree that a 2nd concert would be more successful than the first."

Johnson's plan was simple. He categorized the fare as "Recitations, Conversational remarks, Songs (with words), Cornet Solos, Animal Mimicry, Laughter, Coughing, etc., etc.," which would be "delivered into the mouth of the machine, and subsequently reproduced." He described getting a lot of laughs by trying to sing himself, but he also tried to entice volunteers from the audience or otherwise to take advantage of local talent.[10] Another Edison associate, Professor J. W. S. Arnold, filled half of Chickering Hall in New York City, where his phonograph "told the story of Mary's little lamb" and then, like Johnson's phonograph, rendered a medley of speaking, shouting, and singing. At the end of the evening Arnold distributed strips of used tinfoil, and there was reportedly "a wild scramble for these keepsakes."[11]

These early exhibitions helped establish a formula for Redpath's agents to follow. Redpath himself managed a short season at Irving Hall in New York City, but the typical exhibition was more provincial. In nearby Jersey City, New Jersey, for instance, demonstration rights were owned by Frank Lundy, a journalist, who displayed little polish during exhibitions and who complained bitterly to the company that his territory was always being invaded by others or usurped by Edison's own open-door policy at the Menlo Park laboratory, a short train ride away. Lundy came through Jersey City in mid-June. He gave one exhibition at a Methodist Episcopal church ("admission 25 cents"), and another at Library Hall as part of a concert given by "the ladies of Christ Church." Both programs featured musical performances by community groups as well as explanations and demonstrations of the phonograph. Lundy reportedly "recited to" the machine, various "selections from Shakespeare and Mother Goose's melodies, laughed and sung, and registered the notes of [a] cornet, all of which were faithfully reproduced, to the great delight of the audience, who received pieces of the tin-foil as mementos." But poor Lundy's show on June 20th had been upstaged the day before by a meeting of the Jersey City "Aesthetic Society," which convened to wish one of its members bon voyage. Members of the "best families in Jersey City" as well as "many of the stars of New York literary society" were reportedly received at Mrs. Smith's residence on the eve of her departure for the Continent. For the occasion, one New York journalist brought along a phonograph and occupied part of the evening recording and reproducing laughter and song, as well as a farewell message to Smith, and a certain Miss Groesbeck's "inimitable representation" of a baby crying. Of these recorded cries, "the effect was very amusing," and the journalist "preserved the strip" of foil, saving the material impressions of Groesbeck's "mouth" impressions.[12]

Phonograph exhibitions such as these relied upon a familiar rhetoric of educational merit. Lecturers introduced Edison's machine as an important scientific discovery by giving an explanation of how the phonograph worked and then enacting this explanation with demonstrations of recording and playback.[13] Audiences were edified, and they were entertained. They learned and they enjoyed. Phonograph exhibitions thus reinforced the double message of the lyceum movement in America, sugarcoating education as part of an elaborate ethos of social improvement.

Phonograph exhibitions flirted with the improvement of their audiences in three distinct yet interrelated ways. First, they offered all in attendance the opportunity to participate, at least tacitly, in the progress of technology. Audiences could be up to the minute, apprised of the latest scientific discovery, party to the success of the inventor whom the newspapers were calling the "Wizard of Menlo Park." They were also exposed—playfully and again tacitly—to "good taste." In making their selections for recording and playback, exhibitors made incongruous associations between well-known lines from Shakespeare and well-known lines from Mother Goose, between talented musicians and men like Edward Johnson, between inarticulate animal and baby noises and the articulate sounds of speech. Audiences could draw and maintain their own distinctions, laugh at the appropriate moments, recognize impressions, be "in" on the joke. In the process, they participated in the enactment of cultural hierarchy alluded to above. Finally, the exhibitions elevated the local experience to something much larger and more important. Local audiences heard and saw themselves materially preserved on bumpy strips of tinfoil. Those that met in the meanest church basements were recorded just like audiences in the grand concert halls of New York, Chicago, and New Orleans. Audience members could therefore imagine themselves as part of a modern, educated, tasteful, and recordable community, an "us" (as opposed to "them"), formed with similarly modern, educated, tasteful, and recordable people across the United States. The phonograph exhibition, in other words, offered a democratic vision of "us" and "our" sounds, available to the imagination in some measure because they must have hinted at their opposite: "them" and "theirs."

Of course, this vision came notably vested with cultural hierarchies and a local/global matrix—region/nation, here/elsewhere, local beat/wire story. The familiar practices of public lectures and amusements, the varied contexts within which public speech acts made sense as cultural productions, the enormous, framing tide of newsprint all informed the phonograph's reception. If phonographs were "speaking," their functional subjects remained importantly diffuse among available spoken forms: lectures and orations as well as "remarks," "sayings," recitations, declamations, mimicry, hawking, barking, and so on. The sheer heterogeneity of public speech acts should not be overlooked any more than the diversity of the speakers whose words more and less articulated an American public sphere. The nation that had been declared or voiced into being a century before remained a noisy place.[14]

Actual audience response to the phonograph exhibitions is difficult to judge. Some parts of the country simply were not interested. Mississippi and parts of the South, for example, were far more concerned with the yellow fever epidemic that plagued the region in 1878. Audiences in New Orleans were reportedly disappointed that the machine had to be yelled into in order to reproduce well, and there were other quibbles with the technology once the newspapers had raised expectations to an unrealistic level. Out in rural Louisiana, one exhibitor found that his demonstrations fell flat unless the audience heard all recordings as they were made. Record quality was still so poor that knowing what had been recorded was often necessary for playback to be intelligible. James Redpath spent a good deal of energy consoling exhibitors who failed to make a return on their investments, but he also fielded questions from individuals who, after witnessing exhibitions, wrote to ask if they could secure exhibition rights themselves. To one exhibitor in Brattleboro, Vermont, Redpath wrote sympathetically that "other intelligent districts" had proved as poor a field as Brattleboro, but that great success was to be had in districts where "the population is not more than ordinarily intelligent." Some parts of the country remained untried, while others were pretty well saturated, like parts of Pennsylvania, Wisconsin, and Illinois.[15]

Exhibitors everywhere wrote back to the company for more of the tinfoil that they purchased by the pound. The company kept a "Foil" account open on its books to enter these transactions. Pounds of tinfoil sheets entered into national circulation, arriving in the possession of exhibitors only to be publicly consumed, indented, divided, distributed, collected into private hands, and saved.

Confirming much about the phonograph exhibitions was a "colored burlesque on the phonograph" entitled *Prof. Black's Phunnygraph or Talking Machine.* Frank Hockenbery's skit offers a comment on the phonograph lectures that it lampoons. The term "burlesque" did not then denote striptease, but rather a topical, risqué comedy, full of witticisms pointed at events of the day. The butt of Hockenbery's burlesque were the phonograph demonstrations of 1878. As a "colored" burlesque, *Prof. Black's Phunnygraph* drew much of its "inspiration" from fifty years of minstrel shows. The theatrical season of 1878 had been characterized by a bewildering outbreak of so-called "mammoth minstrel shows," touring troupes of 40 to 60 performers trying to breathe life back into a hackneyed form by finding novelty in numbers.[16] Hockenbery's *Phunnygraph* was probably intended as an interlude in one of these racist pageants, since its concluding stage directions call for a minstrel staple, "moving to half circle, [and as] soon as half circle is struck, begin negro chorus or plantation melody. Minstrel business."

Whatever its exact provenance, *Prof. Black's Phunnygraph* clearly takes aim at phonograph exhibitors like Edward Johnson and Frank Lundy by assuming the form of a lecture "on de Phunnygraph, or Talking Machine, as she am called by de unsophisticated populace." A character adapted as much from medicine shows and Barnum-type attractions as from men like Edison or his friend Professor Arnold, "Professor" Black is the device's "sole inwentor, patenter, manufacturer an' constructioner," although one character makes passing and disparaging mention of a "Billy Addison."

The scenery constituted part of the joke. On stage, a sign announcing the lecture connected the phonograph exhibition to patent medicines:

Admittance
Adults, ...10 cents.
Children, ... ½ dose.

The remainder of the scenery consisted of "a dry goods box large enough to hold three persons" to which had been attached a "sausage-grinder on top with crank," a household funnel, and "slips of white paper to run into [the] grinder to talk on." The lecture that ensued followed the formula of so many Edison Speaking Phonograph Company exhibitions. Professor Black explained how the machine worked and gave a demonstration that went humorously awry. His "phunnygraph," it turned out, was nothing but three people hiding in a box, and it proved impossible for them to remember and accurately repeat the words and noises that the professor shouted into the kitchen utensils on top of the box.

In the circumstances of its demonstration Hockenbery's "phunnygraph" closely resembles the phonograph it mocks. It jabs suggestively at cultural hierarchy even while giving voice to racist cant, since his machine has made its tour "from de highest ranks ob civilization, clean down to de lowest ranks ob dem dat ain't ranked at all." In addition, Professor Black pretends to record talk, recitations, animal mimicry, whistling, and song, just like Edward Johnson and Frank Lundy actually did. There is also a familiar concatenation of imitations: The players imitate African-Americans degradingly, their characters' dialect and malapropisms imitate standard, spoken English, and the "phunnygraph" tries to imitate everything all at once, like Mrs. Groesbeck *and* a baby crying. Finally, the stage directions substitute "white paper" for tinfoil, as if to emphasize that the phonograph forms a kind of writing instrument or to underscore the vexing *blankness* of souvenir tinfoil sheets for their readers.

As Hockenbery's scenery can only hint, it was to the sheets of tinfoil that the language of print culture adhered most freely, becoming confused in the novelty of the phonograph and its records. The material qualities of sheets *as* sheets proposed an equation of writing and recording that proved impossible to work out. Company executives wrote tellingly to each other that a means must be found for "stereotyping" or "electrotyping" records once they were made, drawing their terminology from common printing processes of the day.[17] Meanwhile in the *Gentleman's Magazine,* readers learned about "Edison's imprinted words."[18] And everywhere the term *record* gained a new connotation. Those sheets of tin-foil formed public records in a new way. It was not just that metaphors of writing and printing were being deployed; the literal language of texts was being stretched to encompass these new material forms.

The foil was thus the site of enormous tensions: semantic tension regarding the language of textuality and corresponding semiotic tension regarding the meanings made by or in bumps across the surface of the foil. The habit of collecting scraps of foil made these tensions all the more apparent. Phonograph exhibitors ran through pounds of tinfoil, and audiences scrambled for keepsakes. In their sonic "capture" and later, in their mute evocation of public experience, pieces of tinfoil in private hands formed souvenirs of immense power. They were belongings that vouched for belonging. They were artifacts that vouched for facts. Publicly made and privately held, their material existence offered a demonstrable continuity of private and public memories. Having such a record meant having witnessed and aurally confirmed its public production as a reproduction. It also meant having witnessed its public destruction, as the tinfoil was removed from the phonograph mandrel and given away. And it meant having collected to oneself the potentiality of reproduction, which the foil had ceased to embody in becoming a souvenir. Like the players hiding in Hockenbery's box, scramblers for these keepsakes must have inexactly recollected sounds even as they collected souvenirs.

Susan Stewart's words on what souvenirs accomplish in general are particularly suggestive in this light. She writes,

> Within the development of culture under an exchange economy, the search for authentic experience and, correlatively, the search for the authentic object become critical. . . . We might say that this capacity of objects to serve as traces of authentic experience is, in fact, exemplified by the souvenir. The souvenir distinguishes experiences. We do not need to desire souvenirs of events that are repeatable. Rather we need and desire souvenirs of events that are reportable, events whose materiality has escaped us, events that thereby exist only through the invention of narrative.[19]

The desire for tinfoil must have been partly a desire for authenticity, for what had really transpired. More than any souvenir program, photograph, pencil, or key chain, though, the tinfoil records suggested the authentication of actual sounds that had been shared— shared by a group of listeners, but also shared by the machine, its sensitive and yet entirely insensate body on the table in plain sight at the front of the demonstration hall. Tinfoil offered a new and curious sort of quotation, perhaps, or a way of living with the question of quotation as never before. To put it another way, the tinfoil souvenirs suggested that oral productions might be textually embodied as aural reproductions, rather than as the usual sort of graphical representation, spelled out and wedged between quotation marks on a page. What audiences had witnessed was not the special performance of texts (for example, a written declaration making a nation independent, paper instruments of law making law, canonical texts making a national tradition). Rather, they had witnessed a special textual-ization of performance—quotation somehow made immanent, quotation *marks* of a new sort, which turned (or *re*turned, mechanically, miraculously) into the quotations themselves.

Tinfoil records were souvenirs that offered the renovation of the souvenir as such, hinting at changes to the then normal connections between matter and event, stuff and utterance, text and speech act. The phonograph demonstrations were indeed only reportable, not repeatable, in Stewart's terms—it was devilishly hard to get a tinfoil record back onto the machine once it had been removed, and certainly impossible to reproduce anything from it when it had been ripped up and distributed among the audience. Still, what could be reported about the demonstrations was precisely their repeatability. Narrating the meaning of the tinfoil at breakfast meant testifying to the pending usurpation of that very narrative. The desire for authenticity would finally be consummated, when some now imaginable souvenir spoke for itself. The morning papers promised as much. Tinfoil and newsprint lying side-by-side on the breakfast table served to interrogate that promise, if also to prompt the witness/auditor who owned them and who formed part of their collective, collected subject.

What happened next? In one sense, nothing. Phonograph demonstrators across the country had quickly milked the exhibition cow dry. As Hockenbery's Professor Black put it, the device had been displayed with "profound satisfaction and excess." By 1880 its novelty was gone; the Edison Speaking Phonograph Company soon folded its operations; and James Redpath went on to fresh projects. The phonograph would wait almost a decade for any further development, when it would be joined by a twin, the graphophone, and another sibling, the gramophone, all of them replacing tinfoil sheets with more durable matter, respectively, wax cylinders and shellac disks. One more flurry of public demonstrations ensued before the machines were adapted for musical playback and domesticated as commodities for private ownership. Tinfoil souvenirs went the way of all souvenirs, most of them lost and a very few of them reinvested with meaning as the specific experience of their production faded from memory. Allen Koenigsberg's masterful *Patent History of the Phonograph* comes in a limited second edition (1991) that offers a tiny square of "Original Phonographic Tinfoil (1878)" pasted behind its title page beside notice of copyright.[20] There the tinfoil still speaks mutely of its own power. It and the page to which it is affixed exist in dialogue about the relationships between pages and history, matter and event. The particulars of a specific exhibition hall, a specific phonograph mandrel, and a specific breakfast table are long ago forgotten with regard to any one scrap of foil, and the tiny squares are all unredeemable as recordings except insofar as Koenigsberg's book and others like it narrate their identity. Yet the tiny squares of bumpy foil still vouch for words uttered or sounds made during some fraction of a second in the year 1878. And they vouch in a way that pages, letters, and quotation marks cannot.

In another sense, however, souvenir sheets of foil forever entered the process of discerning textuality and thus printedness. As a means of preservation the tinfoil phonograph helped to raise emphatic questions of loss within which the efficacy and the meaning of print had long been embroiled. Narrowly a matter of words written down, published, and saved, these questions of loss also suggested much broader practices of cultural self-identification, preservation, and interpretation, of "our best poets" saved against "our" uncertain future, for example, and of Onondaga "accents" "preserved" against "their" imminent demise. It was in this larger, us-and-them sense that culture seemed to require self-conscious preservation. National, ethnic, linguistic, and racial identities needed collection, both in the sense of being possessed and of being understood. Yet the tinfoil souvenirs somehow seemed to hint at loss, for they disparaged print by implication and helped to suggest the latter's *in*authenticity as a means of recording and preservation: "Speaking" on paper offered no actual sounds of speech. Yet at the same time, the evident inadequacies of tinfoil records as permanent or indelible inscriptions put the storied, paperless future (which Edison kept vaunting in the press) on hold.

When Edison boasted to the newspapers that his invention would record novels cheaply and thereby ruin the market for books, he reasoned that recordings were printed naturally, by sound waves on tinfoil, rather than laboriously set in type by human compositors. Authors and their audiences would win, even if printers and compositors would lose. But when the inventor's comments reached their readers, set in type of course by compositors, the inventor was quoted as saying that a phonograph record "is not in tpye [*sic*], but in punctures" on tinfoil. The typographical error is exactly that, graphical. It produces an unpronounceable "utterance," which remains viable only on the page, seen and not said. The four individual pieces of type remain insistently, visually, sorts of type. They are emphatically t, p, y, and e, since they do not make the word *type*. Yet the words are also emphatically true; like the vaunted records of the future, they are "not in tpye," because they are in type. There is no way to know for sure whether this was accident or derision on the part of a compositor employed by the *Philadelphia Weekly*—lazy or Luddite? rushed or radical?—and it is unlikely that many readers stumbled or that any noticed their dilemma. A trivial instance, then, but one that neatly captures the verbal-visual status of writing and print that the first phonograph records uncomfortably or unfamiliarly visited. "Behold!" wrote early auditors (instead of "Listen!"), "The voice of the author is heard repeating his own composition." In their rush to celebrate the greater immediacy of a new medium, contemporaries got the sensory apparatus wrong.

Far less trivial than the typography of one word on the future of literature was Moses Coit Tyler's publication of a two-volume version of the American literary past. Only a scattering of colleges or universities taught any American literature when Tyler published his *History of American Literature, 1607–1765* in 1878. He was among the first to put together a coherent narrative of American literary history, and his project helps characterize the cultural status of writing and print at the moment Edison's speaking phonograph was introduced to American audiences. As Tyler himself described them, his two volumes sought to present an exhaustive history of those writings in English by Americans that "have some noteworthy value as literature, and some real significance in the unfolding of the American mind." His method had been to troll for years through libraries, seeking out neglected books, assessing their literary merits, and judging their pertinence to "the scattered voices of the thirteen colonies," which eventually and so importantly, in his view, "blended in one great and resolute utterance." It was a hugely ambitious project, and one he everywhere describes in terms that seem to mix the functions of speech and writing. He wanted to see just where "the first lispings of American literature" took their departure from their "splendid parentage, the written speech of England." Tyler's "lisped literature" and "written speech" confirm the degree to which spoken and written language were integrally, mutually, defining. The substitution of "Behold!" for "Listen!" partook of an ancient tradition still current. Writing in general, and literature in particular, lived and was valued in intricate association with speech acts.[21]

The word *record* encapsulates these points. Tyler begins his first chapter proposing the intellectual history of America: "It is in written words that this people, from the very beginning, have made the most confidential and explicit record of their minds. It is these written records, therefore, that we shall now search for that record." Despite his reputation as a "superb stylist," Tyler's double use of "record" is confusing. Mental records are his broader category, within which written records stand preeminent. Confusion between the two is indicative of tensions surrounding the potentially *national* character of textuality for Tyler and, more generally, the gap he seems anxiously to have sensed between minds and pages, between an author's conception and its written, printed expression. Whether Tyler composed these sentences just before, after, or in light of Edison's records, his diction partakes of the selfsame context.

Not surprisingly perhaps, Tyler took liberties when quoting the literary tradition he established. He modernized spelling and punctuation and considered that it was "no violation of the integrity of quotation" for him to expunge or correct any confusion, "extreme inaccuracy," or "palpable error of the press." Against such liberal quotation practices, and such mistrust of the press,[22] and in keeping with his larger project to collect and preserve an explicitly American tradition, Tyler's "record" is a keyword in the sense that Raymond Williams picked out his *Keywords*. Like Williams's word *culture* (or *class,* or *media*) that is, Tyler's *record* seems to have "acquired meanings in response to the very changes he proposed to analyze."[23] The term cannot be defined without recourse to the very conditions it connotes: the lispings of literature and the construction of tradition. Though obvious in Tyler, the reflexivity of "record" proved fleeting, most likely because playing phonograph records soon gained a broad and habitual appeal.

I am hinting at the power of an unfulfilled desire, part of the cultural construction of any new medium. Phonograph exhibitors promised a new, less mediated inscriptive form; their recordings offered more immediate access to public speech acts than any written or printed instrument ever could. If this promise remained only that in 1878—a promise—it nevertheless provoked repeated and sustained vernacular experiences of the relationships between speech and writing, relationships theorized only later by linguists and philosophers. Breakfast-table conversations must at some level have apprehended what it might mean to "risk death in the body of a signifier," to use one much later parlance, as a saved, savable scrap of voice-indented foil sat beside the periodic and ephemeral newsprint of which it and its saver were the subject.[24] Tinfoil scrap and newsprint squib: They formed a pair of characters in private conversation about public speech. One was new and both were flawed, those flaws apparent in the contrast between them. The rank artificiality of "speaking" on paper was newly party to the tangle of concerns that tinfoil souvenirs helped to adumbrate.

Recorded sound eventually prospered, of course, but the newspapers of 1878 remain the best *record* of its public introduction. Into the circulation of the newsprint and the circuits of the American lyceum entered the touring phonograph exhibitors. With their own modest circuits of mail, of revenue, and of foil they immodestly boosted the phonograph in public, promoting it to, as well as via, newly recordable Americans.

Notes

1. "Phonograph and Its Future," *North American Review* 126 (June 1878): 527–536. One portion of this essay has appeared previously as "First Phonographs: Writing and Reading with Sound," *Biblion* 8 (1999): 3–16. Many people offered generous comments regarding earlier versions that have been vital to its revision. Among these I must particularly acknowledge Jay Grossman's conference comments and Anne Skillion's editorial advice.

2. Kenneth Cmiel, *Democratic Eloquence: The Fight Over Popular Speech in Nineteenth-Century America* (Berkeley: University of California Press, 1990), 177–184, passim.

3. Jon Cruz, *Culture on the Margins: The Black Spiritual and the Rise of American Cultural Interpretation* (Princeton: Princeton University Press, 1999), 3.

4. My speculative argument takes Benedict Anderson and David R. Shumway as points of departure, respectively *Imagined Communities: Reflections on the Origin and Spread of Nationalism,* rev. ed. (London: Verso, 1991) and *Creating American Civilization: A Genealogy of American Literature as an Academic Discipline* (Minneapolis: University of Minnesota Press, 1994). Shumway traces the ideological project of disciplinary formation, notably the verification of American culture in the construction of American literary studies (p. 6); my attention is to *anxieties* attending and thus motivating such a project.

5. *Scientific American* 37 (December 1877): 384.

6. "The Edison Speaking Machine, Exhibition Before Members of Congress," *New York Times* (April 20, 1878), 1:1.

7. This from a pamphlet by Frederick F. Garbit, *The Phonograph and Its Inventor, Thomas Alvah [sic] Edison.* Boston, 1878.

8. Charles F. Horner, *The Life of James Redpath and the Development of the Modern Lyceum* (New York: Barse & Hopkins, 1926), 227, 185.

9. Records of the Edison Speaking Phonograph Company exist at the Edison National Historic Site in West Orange, N.J., and at the Historical Society of Pennsylvania in Philadelphia. Documents from West Orange have been microfilmed and form part of the ongoing *Thomas A. Edison Papers, A Selective Microfilm Edition,* ed. Thomas E. Jeffery et al., 4 parts to date (Bethesda, Md.: University Publications of America). These items are also available as part of the ongoing electronic edition of the Edison Papers; see *http://edison.rutgers.edu.* For the items cited here, like Uriah Painter to Thomas Edison of August 2, 1879 ("milked the cow"), microfilm reel and frame numbers are given in the following form: TAEM 49:316. Documents from Philadelphia form part of the Painter Papers collection and have been cited as such. The company's incorporation papers are TAEM 51:771. The history of the company may be gleaned from volume 4 of *The Papers of Thomas A. Edison, The Wizard of Menlo Park,* ed. Paul B. Israel, Keith A. Nier, and Louis Carlat (Baltimore: Johns Hopkins University Press, 1998). Also see Paul Israel's "The Unknown History of the Tinfoil Phonograph," *NARAS Journal* 8 (1997–1998): 29–42.

10. Johnson to U. H. Painter of January 27th, 1878, in the Painter Papers; Johnson prospectus of February 18, 1878, TAEM 97:623. Both are transcribed and published in volume 4 of the *Papers of Thomas A. Edison*. I am grateful to Paul Israel and the other editors of the Edison Papers for sharing their work in manuscript and for sharing their knowledge of the Painter Papers.

11. "The Phonograph Exhibited: Prof. Arnold's Description of the Machine in Chickering Hall—Various Experiments, with Remarkable Results," *New York Times* (March 24, 1878), 2:5.

12. Lundy's complaint of August 31, 1878, TAEM 19:109; accounts of Jersey City are reported in the *Jersey Journal*, June 13th, 14th, and 21st, and the *Argus*, June 20th.

13. For a discussion of this improving ethos at later demonstrations see Charles Musser, "Photographers of Sound: Howe and the Phonograph, 1890–1896," chapter 3 in *High Class Moving Pictures: Lyman H. Howe and the Forgotten Era of Traveling Exhibition, 1880–1920* (Princeton: Princeton University Press, 1991).

14. I am thinking here of Benedict Anderson (see note 4), as well as Chris Looby, Jay Fliegelman, and Christopher Grasso, respectively *Voicing America: Language, Literary Form, and the Origins of the United States* (Chicago: University of Chicago Press, 1996); and *Declaring Independence: Jefferson, Natural Language, and the Culture of Performance* (Stanford: Stanford University Press, 1993); *A Speaking Aristocracy: Transforming Public Discourse in Eighteenth-Century Connecticut* (Chapel Hill: University of North Carolina Press, 1999). All of these discussions notably address an earlier period, as does Michael Warner's *The Letters of the Republic: Publication and the Public Sphere in Eighteenth-Century America* (Cambridge: Harvard University Press, 1990). For catalyzing imaginaries (above) Anderson has "the fatality of human linguistic diversity" (p. 43).

15. These details from the Painter Papers, Letter books and Treasurer's books of the Edison Speaking Phonograph Company, including Smith to Hubbard of November 23, 1878; Mason to Redpath of November 1, 1878; Cushing to Redpath of July 16, 1878; Redpath to Mason, July 10, 1878.

16. 1878 also saw a bewildering number of versions of *H.M.S. Pinafore* on the American stage. See Robert C. Allen, *Horrible Prettiness: Burlesque and American Culture* (Chapel Hill: University of North Carolina Press, 1991). Frank Hockenbery, *Prof. Black's Phunnygraph; Or Talking Machine, A Colored Burlesque on the Phonograph* (Chicago: T.S. Denison, 1886). An original printed copy of Hockenbery is at the New York Public Library, available elsewhere in microprint as part of the *English and American Drama of the Nineteenth Century* series (New York: Redex Microprint, 1968). Its production history is unknown.

17. Painter Papers Letter books, Cheever to Hubbard of June 10, 1878.

18. Richard Proctor, "The Phonograph or Voice Recorder," *Gentleman's Magazine* 242 (1878): 688–705.

19. *On Longing: Narratives of the Miniature, the Gigantic, the Souvenir, the Collection* [1984] (Durham: Duke University Press, 1993), 133, 135.

20. *A Patent History of the Phonograph, 1877–1912,* rev. ed. (Brooklyn: APM Press, 1991).

21. Moses Coit Tyler, *A History of American Literature, 1607–1765* (New York: Putnam's, 1878), xi–xv, 5, 11. On Tyler and his style, see Kermit Vanderbilt, *American Literature and the Academy: The Roots, Growth, and Maturity of a Profession* (Philadelphia: University of Pennsylvania Press, 1986), 81, 84.

22. An even better example of a similar mistrust of print was the ballad collectors' determination to value unpublished sources above published ones. Francis Child devoted himself to locating manuscript sources and the sources that "still live on the lips of the people," G. L. Kittredge, "Preface" *The English and Scottish Popular Ballads,* ed. Francis James Child, 5 vols. (New York: Dover, 1965), xxvii–xxviii. Childs's work appeared in parts, 1883–1898. American "Folk-Lorists" felt the same way. "Negro" folklorist Alice Mabel Bacon urged, "nothing must come in that we have ever seen in print" (1897); quoted in Cruz, 170. A generation later, the "new bibliography" movement set out to establish editions of canonical "works," salvaged from the (potentially corrupt) printed "texts" of the past.

23. Quoted is Leo Marx from his work on another keyword, "Technology: The Emergence of a Hazardous Concept," *Social Research* 64 (fall 1997): 965–988. See Raymond Williams, *Keywords: A Vocabulary of Culture and Society* (New York: Oxford University Press, 1976).

24. "Risk death" is Jacques Derrida, quoted in Bruce R. Smith, *The Acoustic World of Early Modern England: Attending to the O-Factor* (Chicago: University of Chicago Press, 1999), 11. Stated in the extreme, or too extremely, "All concepts of trace, up to and including Derrida's grammatological ur-writing, are based on Edison's simple idea. The trace preceding all writing, the trace of pure difference still open between reading and writing, is simply a gramophone needle," Friedrich A. Kittler, *Gramophone, Film, Typewriter,* trans. Geoffrey Winthrop-Young and Michael Wutz (Stanford: Stanford University Press, 1999), 33.

R. L. Garner and the Rise of the Edison Phonograph in Evolutionary Philology

Gregory Radick

The era of the compact disc player looks back on the phonograph and associates it with the beginnings of recorded music. For those who used the phonograph when it was new, however, playing music was but one of its functions. The phonograph in those days was best known as the "talking machine." Its inventor, Thomas Edison, forecast roles for the phonograph in, among other things, correspondence, dictation, and talking clocks and dolls.[1] It was only gradually, as Edison and his rivals learned what sold and what did not, that most of these talk-related uses faded from the popular image of the device. A vestige survived into the twentieth century in the famous Victor Talking Machine Company advertisement of a dog listening attentively to a recording of "his master's voice." Now HMV is the name of an international chain of CD shops.

This chapter dwells on events in the early 1890s, when the public identity of the phonograph was still protean. According to the January 1892 *Phonogram,* recent improvements in the design of the phonograph were prompting fresh applications. Each day, wrote the editors, "brings us information of still greater triumphs won by the talking machine." As ever, there were practical triumphs, such as its use in recording telephone calls for later retrieval, or in relieving drill sergeants, train conductors, and singing teachers from the drudgery of repeating themselves. What interests me here is a more exotic triumph then in prospect. The editors reported that an expedition bearing phonographs would soon depart for Africa, to discover whether apes spoke something like a language, and so whether humans were indeed, as the evolutionists had claimed, descended from apelike beings. "If, as scientists have asserted, we are descended from the Simian race, and the conductors of this mission succeed in establishing communication with it and discover a genuine vocabulary, we shall then owe to the phonograph the obligation of teaching us our 'Mother Tongue.'"[2] This philological use of the phonograph neatly reverses the relationship between humans, animals and phonographs later made famous in the Victrola image.

Further details of the expedition followed, under the heading "The Phonograph to Aid in Establishing the Darwinian Theory."[3] Leading the expedition was one "Prof. Garner." R. L. Garner was a Virginian, an evolutionist, a scientific amateur, and an unknown until he began using the phonograph to investigate relations between human talk and simian talk. Garner was heading to western equatorial Africa to work with the simians nearest to humans in physical and mental makeup, the gorillas and chimpanzees. Just as anatomists had revealed how human bodies related to ape bodies, Garner hoped to discover the roots of human language in ape howls and cries, and thus, in his view, to vindicate the still-embattled theory of evolution.[4] Using the phonograph in the zoological gardens of the United States, he had become convinced that monkeys, at least, used words no different in basic structure or function than human words, indeed that the words of the "simian tongue," as he called it, contained the rudiments out of which human languages could have evolved. Now he was going to test his conclusions among the apes. Taking precautions comparable to those biologists take today when studying sharks in the wild, Garner was bringing with him a large metal cage. Seated in his cage deep in an African forest, he hoped to recover, with the aid of the phonograph, language's missing links.

Here I aim to reconstruct the phonograph's brief career as an "aid in establishing the Darwinian theory."[5] I begin with Garner, looking in particular at how he came to use the phonograph to investigate the evolutionary origins of language. Next I turn to the phonograph itself, and the changes that, from the late 1880s, made it attractive for use in Garner's kinds of experiments. The third part of the chapter tracks the astonishing rise to prominence of Garner and his claims, from his first publication on his phonograph experiments in June 1891 to his departure a year later for the French Congo; the fourth part examines the collaboration between Garner and Edison on new, specially modified phonographs. In the fifth part I look more closely at how Garner used the phonograph, distinguishing several different experimental roles, and also drawing attention to Garner's own understanding of how the phonograph worked. I conclude with some reflections on new media and old, and new sciences and old, at the end of the nineteenth century.

Garner

Richard Lynch Garner does not much figure in the historiography of the evolutionary debates, although from the early 1890s until his death in 1920, he was one of the best-known evolutionists in the world.[6] He was born in 1848 in the town of Abingdon, in the

Blue Ridge Highlands of southwestern Virginia.[7] Although the region was poor, rural, and isolated, Abingdon was its commercial and cultural center, and as such grew prosperous and even genteel. Its historic downtown still boasts a number of large and elegant buildings from the mid-nineteenth century, when Abingdon bustled with shopkeepers, bankers, doctors, lawyers, and merchants, the latter trading in salt, wool, flax, livestock, and other products from the surrounding farms, mines, and mills.[8] There were churches, schools, and halls for lectures and theatrical performances.[9] Garner's father, Samuel, supported his large family as, at one time or other, a horse and cattle trader, a mill owner, proprietor of a foundry, and a plough and tin supplier.

He sent Richard to private school from the age of eight to prepare for the ministry. In later life the failed minister recalled an early but silent skepticism about religion. "I was surrounded on all sides with the bulwarks of orthodox religion and any one, old or young, who would have confessed to the sin of unbelief would have been anathematised from every pulpit and fireside in the land."[10] As Garner entered adolescence, hostilities between North and South intensified. When war broke out, Garner's father (who had owned several slaves) and older brothers served in the Confederate army, as did Garner himself, although only in his teens. Fighting at times with Forrest's Third Tennessee Mounted Infantry, later celebrated in the South for its raids on Union territory, Garner served in the battles of Bull's Gap, Cheek's Cross Roads, and Morristown, emerging unharmed, although spending a good deal of time in Union prisons.[11]

After the war, Garner studied from 1865 to 1867 at the Jefferson Academy for Men, in Blountville, Tennessee, taking courses in divinity and medicine. He took up school teaching, a profession he pursued in the region for the next fourteen years. Soon with a small family of his own, Garner supplemented his income as he could. One of his more colorful ventures outside the schoolhouse involved crossing the Missouri and Kansas plains on foot, riding broncos through southern Colorado, fighting Apaches in their northern lands, and capturing and breaking wild ponies to sell in the Texas markets of Sherman and Paris.[12] It was in the course of his work as a teacher, however, that Garner first encountered the scientific controversy over the nature and origins of language.

The occasion was a meeting about the latest developments in phonetics. In the nineteenth century the subject had tremendous intellectual and social cachet. As the goals of mass literacy and efficient communications had spread, so had awareness of the limitations of the conventional alphabet. The phonetic dream was to replace that hodgepodge with a system of simple, accurate, and self-explanatory visual analogs for the sounds of human speech. Just by looking, people would know exactly how to pronounce a word.

Such a system would be a boon to education, especially for the deaf. It would help adults learn a foreign language. It would promote more efficient communication by stamping out regional differences in pronunciation. It would help the imperial nations govern their colonies better, by ensuring swifter learning of the languages of colonizers and colonized alike. As Alexander Melville Bell, the Scottish elocutionist and inventor of a widely acclaimed "physiological" phonetic system (depicting not sounds but the associated positions of the tongue, throat, and other parts of the vocal apparatus) wrote in his 1869 book *Visible Speech,* with "such a medium of self-interpreting letters, the . . . foundation is laid, and the Linguistic Temple of Human Unity may at some time, however distant the day, be raised upon the earth."[13]

Phonetics was Garner's entrée to the debate on the evolutionary origins of language. "As a young school master," he later recalled, "I once attended a Teachers' Institute where I took part in a discussion on phonetics in the course of which I innocently mentioned the sounds of animals as rudiments of speech—This was instantly and violently assailed as rank heresy and I barely escaped being immolated as an infidel." In defending his view, Garner resolved on the spot, as he wrote, "to study those sounds in a methodic manner and try to learn the speech of animals." He claimed to find no precedents for the kind of investigations he aspired to do. Despite the "universal negation and derision" of those around him, he kept at the problem, although the slow progress and "nebulous data" were sources of frustration.[14]

What Garner knew of evolution he probably learned from books, newspapers, magazines, and lectures during the "vogue of Spencer" that swept Gilded-Age America.[15] He read Herbert Spencer, and also Darwin, Tyndall, the anthropologist Rudolph Virchow, and the physicist Hermann Helmholtz. Press reports on a bitter row between the philologists Friedrich Max Müller and William Dwight Whitney in 1875–1876 carried Müller's already famous views to a much wider popular audience, and Garner later recalled Müller among several authorities he had read who insisted, in Müller's famous phrase, that language was "the one great barrier between the brute and man."[16] In the 1880s Garner turned from teaching to business, achieving fitful success in real estate, stock sales and other enterprises.

It was in 1884, at the zoological garden in Cincinnati, that his long doubts about a language barrier between humans and animals were confirmed. After listening for a while to the chatter of a group of monkeys, he found he was able to predict the behavior of a mandrill sharing their cage. Further translation proved elusive, however.[17] Although he continued this research as he could,[18] he soon turned his attention to the origins of writing, in particular to the enigma of the Maya glyphs, which he examined at the Smithsonian whenever business took him to Washington.[19] Attempting in 1888 to advance his studies, Garner asked the curators to furnish him with photographs from several of the collections. His frequent entreaties at last exasperated the Smithsonian officials: "This man is a weariness to my flesh," complained one in an internal memo. "He seems to be pegging away at something, but he never publishes anything."[20]

Garner indeed had a novel scheme in the works. The basic idea was similar to that in Bell's *Visible Speech*. Just as deaf-mutes could learn to speak and understand English through close attention to the outward shapes of the mouths of English speakers, so Garner hoped to learn the language of the Maya through an examination of the mouths of the figures in the Palenque glyphs. "The main feature of those glyphs, by which I was guided, was the outline of the mouth, which the artist had sought to preserve and emphasise at the cost of every other feature," Garner later wrote, "and by this process I found to my satisfaction some ten or twelve sounds or phonetic elements of the speech used by these people; but not knowing the meaning of the sounds in that lost tongue, I did not attempt to verify them."[21]

It was probably on one of his trips to Washington that Garner first saw in the cylinder phonograph a solution to his problems with the monkey language.[22] As he later recalled, "At last came the Edison phonograph. I was being shipwrecked, when this wonderful machine saved me."[23]

The Wax Cylinder Phonograph

Thomas Edison had come upon the idea for his machine in the course of work on telegraph and telephone improvements in his Menlo Park, New Jersey, laboratory in 1877.[24] The first phonographs basically had two parts: a thin metal disc or "diaphragm" with a stylus attached to one side; and a grooved, cylindrical drum, wrapped with a sheet of tinfoil, and mounted on a crankshaft. Unlike gramophones and the later phonographs, the pioneer phonographs could record as well as reproduce sounds. To record, the user aimed the sound to be recorded (sometimes through a horn or "trumpet") at the diaphragm while working the hand crank. As the diaphragm vibrated, the sound etched its pattern via the stylus onto the rotating tinfoil. The result was a record of the sound. To reproduce the recorded sound, the user fitted a smoother stylus and once again cranked the handle, letting the stylus ride over the tinfoil record as it rotated. Now the record set the stylus in motion, which set the diaphragm in motion, which set the surrounding air in motion so as to recreate the original sounds. At first the mere existence of such a machine was cause for astonishment. But the sound quality was poor, and the records themselves wore out after just a few plays. Soon the novelty wore thin, even for Edison, who rapidly moved on.[25]

For the next ten years Edison turned his prodigious energies to the more lucrative project of manufacturing incandescent lamps and distributing electric power.[26] In 1887 he returned to the phonograph with a literal vengeance. His competitors had begun (behind his back, as he saw it) to patent improved versions of the device. Out of Edison's belated burst of enthusiasm came the so-called "perfected" phonograph. It was notable for a number of innovations, but two in particular. First, in place of the humble hand crank of the early phonographs, there was a heavy electric motor, often powered by a lead-acid battery and regulated by a rather cumbersome governor. Second, in place of the tinfoil sheet, wrapped around a mobile drum, there was a wax cylinder, slipped onto a stationary spindle or "mandrel." Wax recordings made at more constant speeds sounded better, recorded longer, and endured more robustly—hundreds of plays were now possible.[27] "Here is a pretty little machine," wrote one admirer, "a little bundle of iron nerves and sinews imbued through electricity with the great principle of life, 'the soul of the world,' chained by Edison's genius and forced to obey our behests."[28] When *Punch* brought Michael Faraday back from the grave in 1891 to pronounce his pleasure at Science's progress, the perfected phonograph had center stage, surrounded by the paraphernalia of cylinder containers, an enormous battery, and other examples of Edisoniana, including a chart of telegraph cables, a telephone, and a microphone (see figure 8.1).[29]

Despite massive publicity, the improved phonograph remained a commercial underachiever. In Europe and Britain, public expositions and demonstrations were really the only places where most people encountered the phonograph.[30] In the United States phonographs were available for rental or purchase in several regions; but they were just too expensive for most people to have in the home.[31] Edison was keen to see phonographs put to practical use in offices, for taking dictation or sending messages.[32] Although the phonograph met with only limited success in this regard,[33] Washington proved a welcome exception. "During the session of Congress between fifty and sixty machines were used in the Capitol alone, by Senators, representatives, officials, etc.," reported the April 1891 *Phonogram*. "Every department of the Government is now supplied. In the homes of men of wealth and culture the phonograph is already a fixture."[34] Outside those homes the phonograph was slowly becoming a fixture as well, in railroad stations, drug stores, saloons, and even new "phonograph parlors." These sometimes unsavory places began to

Figure 8.1   The Edison cylinder phonograph celebrated in *Punch*. Illustration from *Punch* 100 (27 June 1891): 309. Courtesy of Widener Library, Harvard University.

feature the forerunners of the jukebox, the nickel-in-the-slot phonograph, where a nickel bought a few minutes of the latest John Philip Sousa march or Stephen Foster ditty.[35]

A new scientific use for the phonograph had been announced the previous spring by Jesse Walter Fewkes, one of a new, professional breed of anthropologist, working full-time for the United States government under the auspices of the Bureau of American Ethnology at the Smithsonian.[36] Writing in *Science* "on the use of the phonograph in the study of the languages of the American Indians," Fewkes described "experiments" (his term) he had made with the phonograph in preserving the speech, songs, folk-lore, and rituals of the Passamaquoddy Indians of Maine. "What specimens are to the naturalist in describing genera and species, or what sections are to the histologist in the study of cellular structure, the cylinders made on the phonograph are to the student of language," he wrote. No longer was the ethnologist hostage to an imperfect ear, a faulty memory, hopeless systems of notation, limited time, or, indeed, the extinction of the speakers of a language. "There are stories, rituals, songs, even the remnants of languages which once extended over great States, which are now known only to a few persons," wrote Fewkes. "The phonograph renders it practicable for us to indelibly fix their languages, and preserve them for future time after they become extinct or their idiom is greatly modified or wholly changed."[37]

Was Garner aware of Fewkes's innovations? No doubt they were much discussed among the anthropologists Garner was pestering for access to the Maya glyphs. Whatever he knew about Fewkes, Garner conducted his first phonograph experiments with monkeys at the Smithsonian in September 1890. In November he sent a phonogram to the anatomist Frank Baker, acting director of the latest addition to the Smithsonian imperium, the National Zoological Park, proposing some further experiments. Not that Baker understood this immediately. He and his staff gave the posted cylinder a spin on the office graphophone (a commercial rival of the phonograph, produced by telephone inventor Alexander Graham Bell's Volta Laboratory in Washington).[38] They tried the cylinder one way round, then the other, but unintelligible noise was the result each time. As Baker later wrote to Garner, "I came to the conclusion that it represented an entire carnivore house in full blast about feeding time." No letter of explanation or instruction came. At last someone in the office suggested that the mystery cylinder might need to spin at a speed outside the range of the office machine. Baker went to the headquarters of the Columbia Graphophone Company to find the right machine. Several attempts with various models gave equally poor results. At last one of the older machines, faster than its descendants, was tried, and the roars and bellows became articulate speech. Baker wrote back encouragingly in December—although he advised Garner to take more care in his future attempts to communicate by cylinder.[39]

Early in 1891 Garner was back in the monkey houses of Washington, armed with a new piece of apparatus, and in April, the *Phonogram* carried a small notice about "novel experiments with the phonograph":

A learned professor at the Smithsonian Institute has recorded on the phonographic cylinder the chatter of monkeys; and, after careful practice of the sounds, he finds that, by repeating them, he can make himself understood by the animals. Sounds expressive of fear, cold, hunger, and other sensations common to the human race and its four-footed brethren, have already been recognized. It is asserted that there are forty different sounds in the Simian vocabulary.[40]

Publicizing Garner's Phonograph Experiments

Garner gave a full account of his research in "The Simian Tongue," published in the June 1891 *New Review*. As he described it, his basic method was to record the utterances of monkeys and then play the records back to the monkeys, watching their responses. The phonograph enabled Garner for the first time to repeat back to the monkeys their exact utterances, under controlled conditions, and also to slow down and analyze their forbiddingly rapid speech. On the basis of his phonographic work, he wrote, "I am willing to incur the ridicule of the wise and the sneer of bigots, and assert that 'articulate speech' prevails among the lower primates, and that their speech contains the rudiments from which the tongues of mankind could easily develop; and to me it seems quite possible to find proofs to show that such is the origin of human speech."[41]

Garner's article attracted immediate attention. The versifiers at the English satirical weekly *Punch* responded with a poem, "Simian Talk:"

> PROFESSOR GARNERS, in the *New Review*
> Tells us that "Apes can talk." *That's*
>   nothing new;
> Reading much "Simian" literary rot,
> One only wishes that our "Apes" could *not!*[42]

Among the most famous readers of Garner's article was the Harvard philosopher and psychologist William James. In his copy of his recently published *Principles of Psychology,* next to the sentence "Man is known again as 'the talking animal'; and language is assuredly a capital distinction between man and brute," James inked in a note: "Cf. R. L. Garner, the Simian Tongue, (New Review, June 1891)."[43] The English *Spectator* celebrated Garner's article under the heading "Apes and Men." For the *Spectator,* Garner's phonograph was an emblem, not of the ties that bound apes and men, but of the free will that scientists exercised in examining the laws of nature, laws that governed apes and all else nonhuman.

Where Garner as the human investigator had freely chosen how to direct his attention in conducting his experiments, his simian subjects attended only to that which excited them and their desires. Hence the "whole series of experiments, and the wonderful instrument which rendered them possible, are a triumph of this one great fundamental root of all science, the act of voluntary attention."[44]

Why proclaim the discovery of the simian tongue in the *New Review*? In several respects, it was not an obvious choice. While old lions such as Thomas Huxley, Herbert Spencer, and Alfred Russel Wallace still held forth on the nature of life and evolution in the *Nineteenth Century* and other nonspecialist journals, empirical work in the natural sciences now appeared largely in journals devoted to the natural sciences, indeed to particular natural sciences.[45] Even among the nonspecialist journals, the *New Review* was neither especially successful nor well known for its coverage of science. It was founded in 1889 by Archibald Grove, a young Liberal MP who hoped the review's combination of low price and serious content would fill a gap in the periodical market. While articles on science and technology appeared from time to time, politics and the arts were the mainstays of the *New Review*. Garner's articles appeared alongside work by Henry James, Emile Zola, and a host of lesser but still considerable Victorian lights, such as Frederic Farrar, the novelists Grant Allen, Olive Schreiner, and Eliza Lynn Linton, and the mythographer Andrew Lang. But sales were stubbornly low. Rather than filling a gap, the *New Review* seems instead to have fallen into one, being neither as high-caliber as the more expensive intellectual journals, nor as inexpensive as the low-caliber illustrated magazines.[46]

The placing of "The Simian Tongue" in the *New Review* was probably the work of Samuel Sidney McClure, Garner's literary agent. As proprietor of a major literary syndicate, the New York–based McClure in the early 1890s was always on the lookout for fresh writing talent. He had excellent contacts in the English publishing world, and traveled constantly between England and the United States, negotiating terms all the while.[47] His antennas for what would sell were remarkably good; and he seems to have become Garner's agent from early days.[48] It was McClure who turned the obscure Virginian into the talked-about "Professor Garner." The list of outlets where McClure flogged Garner's writings reads like an atlas to a vast newsprint continent: the *Atlanta Constitution;* the *St. Louis Post Dispatch;* the *Detroit Tribune;* the *Minneapolis Tribune;* the *Chicago Tribune;* the *Boston Herald;* the *Philadelphia Inquirer;* the *Montreal Star;* the *Montgomery Advertiser;* the *Cleveland Leader;* the *Louisville Courier Journal;* the *Washington Post;* the *Albany Telegraph;* the *Buffalo Courier;* the *Toledo Blade;* the *Kansas City Times;* the *New Orleans Picayune;* the *Chattanooga Times;* the *Richmond Dispatch;* the *Syracuse Herald;* the *Springfield Republican;* the

Cleveland Plain Dealer.[49] The market was huge, and the well-connected McClure knew how to exploit it. For McClure, Garner represented an investment worth cultivating, a colorful scientist and (under McClure's guidance) soon-to-be explorer who could write vivid prose about his studies in the controversial field of human origins.

In large part thanks to McClure's canny promotional efforts, Garner became news. Shortly after June 1891, according to a reporter for the *New York World,* "the entire scientific world knew of [Garner's] doings," and before long, the "Simian tongue, Mr. Garner's name for his discovery, ha[d] gone all over the world and . . . been written about in every known language save in the simian tongue itself."[50] An exaggeration to be sure; but Garner was being listened to and taken seriously. Barnett Phillips, a reporter sent to Central Park to profile Garner for *Harper's Weekly,* described Garner's phonograph as opening a new route into the prehuman past. "Farther back than man, Mr. Garner is looking for the first understood syllable," wrote Phillips. "If we accept the Darwinian theory, and should Mr. Garner's life work be accomplished, the advocates of evolution will have found a new and strong argument in their favor" (see figure 8.2).[51] Introducing Garner at a meeting of the Nineteenth Century Club in the assembly rooms of Madison Square Garden the following February, the club's president remarked, in the words of a reporter for *Scientific American,* "that Darwin's 'Descent of Man' was the most important scientific work since the 'Principia' of Newton, and that Mr. Garner's brilliant researches were well calculated to sustain the views introduced by Darwin."[52]

Garner followed up "The Simian Tongue" with two more installments in the *New Review,* in November 1891 and February 1892 respectively.[53] He published a number of other articles about his research, in monthlies such as the *Forum* and the *Cosmopolitan*—successful American counterparts to the struggling *New Review.*[54] At McClure's suggestion, Garner wove these articles together, along with much new material, into a book, *The Speech of Monkeys,* published in the summer of 1892 in New York and London by Charles L. Webster and William Heinemann, respectively. The advances came to about $400. This was all the money Garner earned from the book, in large part because Webster went out of business shortly thereafter.[55] In the first half, Garner described his phonographic and psychological investigations among a cast of simians with names such as Jokes, Jennie, Banquo, Dago, McGinty, Pedro, Puck, Darwin ("a quiet, sedate, and thoughtful little monk, whose grey hair and beard gave him quite a venerable aspect"), Mickie, Nemo, Dodo, Nigger, Nellie, Dolly Varden, and Uncle Rhemus.[56] Many of these, he wrote, were his "little friends."[57] Indeed, he "regarded the task of learning the speech of a monkey as very much the same as learning that of some strange race of

Figure 8.2   R. L. Garner with phonograph in Central Park, New York City, and (as imagined by the artist) in West Africa. Illustration by F. S. Church, from *Harper's Weekly* 35 (26 December 1891): 1036. Courtesy of Widener Library, Harvard University.

mankind, more difficult in the degree of its inferiority, but less in volume."[58] The second half of the book was more speculative and theoretical, relating Garner's views on the nature and evolution of speech, thought and life, with remarks on the phonograph and other instruments used in his investigations.

The Speech of Monkeys was widely reviewed, and it was soon translated into German and Russian.[59] The Spectator reviewer emphasized the importance of the manipulative possibilities the phonograph afforded: "By recording the monkey notes on the drum, and then spinning the machine at a slow rate, the sounds are analysed, and modulations detected, and vowel sounds resolved, in a way hitherto impossible."[60] Other reviewers praised the use of the phonograph but lamented Garner's interpretations of his phonographic results. Writing in the Dial, the American psychologist Joseph Jastrow commented: "The author had the happy idea of studying the chatterings of monkeys by recording them in a phonograph . . . but it was one thing to have a good idea, and another to be possessed of the proper ingenuity, patience, and scientific habits, to carry it out."[61] Though he complained of Garner's illogical conclusions and misjudged emphases, E. P. Evans enthused about the prospects of phonographic inquiry. "Mr. Garner's superiority to his predecessors in this department of linguistic research consists in the greater excellence of his material rather than of his mental equipment. The possession of the phonograph alone gives him an immense advantage in this respect, by enabling him to record and to repeat the utterances of monkeys with perfect accuracy."[62] Evans was particularly eager to learn what would come of Garner's work in Africa with "this scientific weapon of phonetic precision,"[63] which "may yet render as valuable service to philology by extending the field of linguistic research as the microscope has rendered to medicine, and especially to bacteriology."[64]

The phonograph was at the heart of the proposed expedition. Using the instrument, Garner expected to learn the tongues of the anthropoid apes, in much the way he had learned the tongues of their simpler kin. But he promised to do much more. He would also record and analyze the tongues of the tribesmen, and compare these with each other and the ape tongues. "Granted that I have got to the bottom of monkey talk, my task would be but half accomplished," Garner told the Harper's reporter, Phillips. "I will have but forged a single link in the chain. I want another. I propose taking down the speech of the lowest specimens of the human race—the pygmies, the Bushmen . . . the Hottentot cluck and click." Garner continued: "If there be family resemblance, structural relationship, between the Rhesus monkey, the chimpanzee, and the lower grades of humanity, there may be correlation of speech, philological kinship, and then—and then—the origin of man's talk might be found."[65] Garner also planned to study how speech developed in individual apes and "natives." Through daily recordings, he would chart this development with precision, allowing him, for example, "to ascertain whether or not the development of consonant sounds follows the same law with them [apes and natives] as with children of the Caucasian race."[66] As well as translating ape words, he would attempt to teach the apes new words, for if "it can be shown that these animals can acquire new sounds and ascribe to them new meanings, it will indicate a capacity which has always been denied them."[67]

Garner and Edison in Collaboration

Among the most important backers of Garner's phonographic expedition was Edison himself.[68] As early as the summer of 1891, letters from Garner to Edison had broached the desirability of certain minor modifications to the phonograph, the possibility of a trip to Africa for further phonographic research among the gorillas and chimpanzees, and concerns about costs and proprietary rights. ("On my return I should want to use the phonograph in lectures, to deliver to an audience the original sounds secured in the wilds of Africa. . . . I should not want to be barred the use of it, whether by any territorial limits or by reason of prices demanded for it.")[69]

Edison already knew of Garner's work, and in a sense had anticipated it. In 1890 the inventor had set down his notes for *Progress,* a Jules Verne–esque novel set in the future. Among the innovations in manufacture, transportation, and warfare Edison envisioned was an "experimental station of the international Darwinian Society at Para on the Amazona," where there were "a great number of educated beings derived from the interbreeding and assiduous cultivation and care of the higher anthropoid apes. Two species being of the eleventh generation were capable of conversing in English."[70] Edison's personal secretary, Alfred Ord Tate, wrote to Garner at the end of August 1891 that Edison had been following Garner's work with interest, and that, provided the Edison United Phonograph Company (which had rights to African distribution) gave its blessing, "Mr. Edison thinks that he will be able to fit you out with some good apparatus by the time you are ready to go abroad. He can provide you with a phonograph that will be twenty times more sensitive and accurate than any yet made . . . [with which] it will be possible to take a fifteen minute continuous record."[71]

Garner obtained the permission of Edison United, and he wrote back to Edison in December with design suggestions for the modified phonograph. Garner's most pressing problem was with simultaneous recording and reproducing. "Very often when I am repeating the cylinder to [the monkeys], one or more of them will respond at different parts of the cylinder," he explained in a letter. "I have been unable to make records of these responses. Using two phonographs I have to labor under the difficulty of either recording the sounds of the first phonograph or lose entirely the sounds of the monkey." He proposed a solution: a phonograph with two mandrels on it instead of one, with both mandrels sharing a single horn, but with one mandrel dedicated to recording, and the other to reproducing. According to Garner, this double-spindle phonograph would not suffer the same problems as two phonographs working together, because the sound waves would pass each other coming from opposite directions, and so would not interfere. Garner included diagrams of the proposed device (see figure 8.3). He also asked that the new phonograph run on a kind of clockwork wind-up spring mechanism, which would be much lighter and more efficient than batteries. Garner's letter concluded with a rather impolitic postscript: "Can you not construct a cylinder similar to those used on the graphophone [of Edison's rival Bell]? Something to stand rougher usage and be less in weight?" Above the letter Edison scrawled: "Say to him that he can come to the lab & I will give him an expert for a day to experiment & learn just what he wants."[72]

That same month, a New York paper ran a vivid description of Garner at work at the zoological garden in New York's Central Park. The problems to do with simultaneous recording and reproducing were just some of a series of obstacles to conducting the phonograph experiments:

> [Garner] attaches a new wax cylinder to the machine and then speaks into the trumpet, saying "Cylinder number one, New York, December 9, 1891." This is, of course, recorded on the wax, and will always be repeated when put into the phonograph.
>
> His plan is to collect monkey words, noting their apparent meaning. By this time one Park policeman and two or three keepers are standing spellbound in the monkey house, and when the experimenter calls for something eatable for the simians all hands hurry to help him. Some pieces of carrot are brought, carefully concealed in paper. The monkeys must not see them until the machine is set going and the trumpet is pointed in their direction.
>
> Mr. Garner bites off a little bit of carrot from a slice and holds it up. "Wh-u-u-h!" he says. A little chap darts forward and makes the same sound, getting the carrot as a reward.
>
> Did you ever try to get a monkey to talk into a phonograph? Don't ever attempt it unless you have lots of time. The little scamps utter their cries freely enough, but the shrewdest of lawyers could not guard himself more carefully against going on record. They won't talk into the trumpet. When they talk into it it is when the wheels aren't

moving, so that their remarks are lost. The experimenter's patience is stupendous. He moves the phonograph and the cumbersome pedestals of boxes and ladders on which it rests half a dozen times over to some promising monkey whose garrulity he thinks he can catch on the fly. . . . The experiments were brought to a close at about half-past ten o'clock [A.M.]. The work had been going on for three hours, but so absorbed and interested were we all that it did not seem half as long. Six wax cylinders were loaded altogether, and as this was all Mr. Garner had, proceedings were discontinued necessarily. No further work could have been done anyway, as a big crowd began to swarm into the monkey house, the report having got out that some exciting things were going on there.[73]

Edison and his assistants set to work on Garner's request for a modified phonograph. Inventor and invention both were now based in the northern New Jersey town of West Orange. Here, almost next door to each other, were the Edison laboratory complex and the Edison Phonograph Works.[74] There was much interest at the time in combining phonographs and telephones. Edison had invented the carbon button microphone that made Bell's telephone workable, and he held numerous patents relating to the telephone.[75] As we have seen, the January 1892 *Phonogram* described the use of the phonograph as (we would now say) a telephone answering machine.[76]

Garner had been planning to take telephones with him all along; but around this time he too became excited about the prospects of a telephonic phonograph. In mid-February, according to one account, he told his audience at Madison Square Garden of a "fine phonograph with telephone attachment . . . being constructed specially by Mr. Edison for Mr. Garner."[77] There was no mention of the clockwork double-mandrel phonograph. Some time later Garner described the new telephonic phonograph in detail:

I shall have attached to the diaphragm of my phonograph [a] telephone wire, which can easily be carried to any point in the forest, within a few miles of my cage, at which point will be attached another telephone concealed in a tin cone, all of which will be painted green, in order that it may be easily concealed in the leaves or moss. Having arranged this at a suitable point, and placed in front of it some bait, effigy, mirror, or other inducement, the place will be watched with diligence, and when an ape may be found within a proper distance if he should utter any sounds they will be transmitted over the telephone wire to the phonograph, where they will be recorded.[78]

It is not clear whether the telephonic phonograph was considered an alternative or an auxiliary to the clockwork double-mandrel phonograph. Although no mention of the latter appears after December, there are at least two reasons for believing that, for a time, both devices were being discussed. First, using a telephone cable to increase the distance

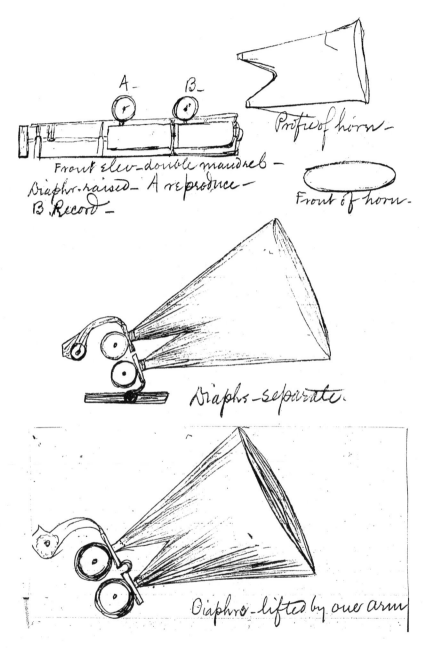

Figure 8.3   R. L. Garner's designs for a double-spindle phonograph. Sketches in Garner's letter to Thomas Edison, 21 December 1891, TAEM 131: 868–870. Courtesy of the Thomas A. Edison Papers, Rutgers University.

between the recording phonograph and the recorded utterance would not obviously have solved the problems of simultaneous playback and recording. Second, when, in April 1892, the secretary of Edison United wrote to Edison for some idea of how much they should charge Garner for the new machine, Edison wrote at the bottom: "Mr. Ott [John T. Ott, probably the expert assigned to Garner] thinks that the apparatus proposed would be uncertain & there is no assurance that the thing will work satisfactorily if it is done. He advises two phonographs." Here the "apparatus proposed" seems to be the double-cylinder phonograph.[79]

However long the telephonic and clockwork double-mandrel phonographs survived as designs, there is no evidence that either device was built before Garner left for Africa in July 1892. Garner put the blame on greed. In *The Speech of Monkeys,* sent to the publishers in early June, he praised the phonograph and its scientific potential, but then complained: "In many ways the use of this machine is so hampered by the avarice of men as to lessen its value as an aid to scientific research, and the Letters Patent under which it is protected preclude all competition and prevent improvements."[80]

The Phonograph as Experimental Apparatus

What, exactly, did Garner do with the phonograph?[81] Often he used it to inspect the phenomena he was studying more closely. Using the phonograph, Garner was able to fix the aural blur of monkey utterances, slow them down, and analyze them. In his words, "to magnify the sounds as I have shown it can be done, allows you to inspect them, as it were, under the microscope."[82] He reported that playing cylinders at speeds much lower than the speed of recording made it possible to "detect the slightest shades of modulation . . . [and] the slightest variation of tension in the vocal chords."[83] Garner magnified the features of monkey utterances still further by using two phonographs in conjunction. While phonograph A repeated a monkey utterance at reduced speed, phonograph B recorded that slowed-down utterance. Then the B-record was itself played at a reduced speed.[84]

In this way the phonograph indeed functioned something like a microscope: It compensated for impoverished senses by putting the world into sharper focus. "The rapidity with which these creatures utter their speech," Garner wrote, "is so great that only such ears as theirs can detect [the] very slight inflections" upon which meaning depended.[85] Of the Rhesus tongue, he claimed: "The 'u' sound is about the same as in the Capuchin word, but on close examination with the phonograph it appears to be uttered in five syllables very slightly separated, while the ear detects only two."[86] If there was indeed more to monkey utterances than met the ear, such as nuances of meaning or subtle phonetic features, and if these hitherto undetectable features did indeed show some kind of continuity with the roots of human speech, then the phonograph was bringing new facts to the debate about the evolutionary origins of language, as Garner claimed.

In some cases Garner did more than just inspect phenomena with the phonograph. He also used to it to purify phenomena. The phonograph enabled Garner to repeat monkey utterances back to monkeys with hitherto unobtainable precision. Furthermore, he was able to screen out or suppress the visual and even (he believed) psychic cues that accompanied the utterances of live monkeys, and so to determine with certainty whether monkeys really did respond to the utterances as such.[87] Typical of his reports is the one on Dago, in Chicago. According to Garner, one of the utterances recorded while Dago was sitting by a window seemed to be a complaint about bad weather. Garner continued: "This opinion was confirmed by the fact that on a later occasion, when I repeated the record to him, the weather was fair; but when the machine repeated those sounds which he had uttered at the window on the day of the storm, it would cause him to turn away and look out of the window."[88] Garner also wrote of the value of the phonograph in removing unwanted features of a sound. By playing a cylinder in reverse, he reported, he was better able to isolate a sound in its fundamental state. "By this means we eliminate all familiar intonation, and disassociate it from any meaning which will sway the mind, and in this way it can be studied to advantage."[89]

Looking again to the emblematic devices of the scientific revolution, the phonograph in these instances functioned less like a microscope than an air pump, enabling intervention in, rather than mere inspection of, the course of nature. But note too an important difference between the phonograph and the air pump. The air pump was located in, indeed helped to define, a specific site, the laboratory. Garner had no phonographic laboratory. Garner and phonograph went where the monkeys were. And monkeys, it seems, were just about everywhere: on the street corners where immigrants played hand-organs, on the stages of traveling shows, in the homes of the well-to-do, and in the ever-burgeoning zoological parks and gardens.

One of the most intriguing uses to which Garner put the phonograph involves neither inspection nor purification so much as the creation of phenomena. Garner regarded as some of his most important findings phenomena that, in a rather straightforward way, he created, by manipulating the sounds he had recorded. Here is a sample of Garner the creator of phenomena:

> By the aid of the phonograph I have been able to analyse the vowel sounds of human speech, which I find to be compound, and some of them contain as many as three distinct syllables of unlike sounds. From the vowel basis I have succeeded in developing certain consonant elements . . . from which I have deduced the belief that the most complex sounds of consonants are developed from the simple vowel basis, somewhat like chemical compounds result from the union of simple elements. Without describing in detail the results, I shall mention some simple experiments which have given me some very strange phenomena.[90]

What is described here is something like this. Garner claimed that monkey utterances were composed almost entirely of vowels. Playing recordings of human vowels at different speeds and so on, he had found, first of all, that these vowels could be broken down to still simpler vowels (of the sort one found in monkey utterances); and second, that the recorded vowels could be manipulated to produce the more complex consonants found only in human utterances. Here we have the use of apparatus to create phenomena that had never before existed. No one had ever noticed the kinds of regularities Garner was producing with his cylinders and phonograph. Moreover, for Garner, these and similarly created phonographic phenomena were *evolutionary* phenomena: They revealed the evolutionary continuity between human speech and monkey speech.[91]

What guarantee was there that the phonograph manipulations legitimately magnified, purified, and extended the deliverances of nature? How could one be sure that the subtleties "revealed" in a slowed-down recording of an utterance, say, were out there in nature waiting to be discovered, and not meaningless artifacts of the process of slowing down? Garner needed a theory of the phonograph to answer such skeptical questions. In *The Speech of Monkeys,* a fair number of pages expound his theory in detail. "At a given rate of speed," Garner wrote in a discussion of his slowing-down technique,

> I have taken the record of certain sounds made by a monkey, and by reducing the rate of speed from two hundred revolutions per minute to forty, it can be seen that I increased the intervals between what is called the sound waves and magnified the wave itself fivefold, at the same time reducing the pitch in like degree. . . . In this process all parts of the sound are magnified alike in all directions, so that instead of obtaining five times the length, as it were, of the sound unit or interval, we obtain the cube of five times the normal length of every unit of the sound. . . . From the constant relations of parts and their uniform augmentations under this treatment, it has suggested to my mind the idea that all sounds have definite geometrical outlines, and as we change the magnitude without changing the form of the sound, I shall describe this constancy of form by the term contour.[92]

Notice the emphasis on the *preservation* of features present in the original recording: "All parts are magnified alike in all directions," "constant relations of parts," "uniform augmentations under this treatment," "we change the magnitude without changing the form." Garner presented a theory of the phonograph that is at the same time a defense of his uses of the phonograph to find out about monkey speech and its relations to human speech. According to Garner, phonograph manipulations of recordings introduced no distortions. He even reported finding this conclusion confirmed in the visual displays of the phoneidoscope—a simple device, about as old as the phonograph itself, in which different sounds caused different colored patterns to form on a soapy film.[93]

New Media and New Sciences

Pulling back from the details of Garner's forgotten experiments, it is time to ask how an awareness of them alters our understanding of the phonograph as a medium. In conclusion I want to explore two possibilities. The first is that Garner's transformation into a celebrated evolutionist, and the phonograph's transformation into an experimental tool, depended as much on developments in the dominant medium of the era, print, as on developments in the phonograph. The second is that the emergence of the phonograph as a medium of communication between humans and simians signposted a science of the animal mind quite different from the one that took hold in the late nineteenth and early twentieth centuries. Let me consider each of these possibilities in turn.

Elsewhere in this volume, Lisa Gitelman argues that the scraps of embossed tinfoil that people carried off from early phonograph exhibitions were understood to be texts of a new kind. She highlights the ways in which the recorded word and the printed word came into competition with one another in the late nineteenth century. But she acknowledges that there were other, complementary sorts of relationship. Garner's story shows concretely how intertwined, indeed mutually enabling, the old and new media could be. What Gitelman calls "the additional subjectivities of late-century literacy" and "the changing political economy of publishing" are helpful pointers.

Additional subjectivities entered the world of the literate in the late nineteenth century in part because new kinds of people were being taught to read. As we have seen, Garner came upon the problem of language origin through his work as a schoolteacher. He made a living teaching students how to read, in an age when professional educators were growing increasingly enthusiastic about phonetics. It was at a teachers' meeting on phonetics that Garner first found himself defending the view that human speech was continuous with the utterances of animals. Out of this encounter came the research program he would later realize with the phonograph.

Changes in publishing stimulated the realization of that program as much as changes in pedagogy stimulated its origin. Garner was not a rich man, and he does not seem to have held a conventional job while pursuing his experimental studies with the phonograph. In large part he funded his experiments through writing about them; and he managed to do this despite a lack of contacts in the world of academic science. Behind Garner's publishing success, and the fast rise in his public reputation as a scientist, was S. S. McClure, who touted Garner and his writings to newspapers and periodicals throughout the United States and beyond. McClure was one of the architects of what has been called the "cheap-magazine revolution of the nineties," when vast amounts of high quality, general interest reading came within financial reach of a large population bent on self-improvement.[94] McClure with his publishing ventures was as much an innovator as Garner with his phonographic ones. Old media and new thus helped one another to flourish—for a time.

Here I have told only of the rise of the Edison phonograph in evolutionary philology, not of its fall. Even the barest outline of the latter, however, will suffice to indicate that it too depended in part on changes in the political economy of publishing.[95] Leaving aside Garner's failure to procure a working phonograph during his time in Africa, the main source of trouble after the expedition was an English six-penny weekly called *Truth,* devoted to high-society gossip and the exposure of frauds in public life. Between 1894 and 1896, the owner-editor of *Truth,* Henry Labouchere, headed up a campaign to discredit Garner. By and large, the campaign was a success. The simian tongue, and the techniques used to discover it, virtually disappeared from scientific discussion, victims of the personality-focused, scandal-mongering "new journalism" of the age at its most vicious.[96]

Turning from old and new media to old and new sciences, the important point is that Garner and his phonograph experiments rose to prominence at a crucial moment in the history of the science of animal minds. The 1870s, 1880s, and 1890s saw far-reaching debate over the methods proper to such a science. Garner's approach combined empirical rigor with a high estimation of the mental abilities of animals. Through much of the twentieth century, of course, the dominant approach was quite the opposite. To be objective about animal minds became, in the first instance, to be skeptical about animals having humanlike mental powers, especially the power of reason. Historians often date such skepticism to 1892. That was the year Conwy Lloyd Morgan, a little-known science lecturer in the English provinces, introduced a now famous rule of method. According to "Morgan's canon," as the rule came to be called, the truly scientific observer explained animal actions whenever possible in terms of lower mental powers, such as trial-and-error learning, rather than higher ones, such as reason. Comparative psychology in the twentieth century was conducted along canonical lines.

But 1892 was also the year that Garner set off for the Congo, determined to bring back phonographic evidence of ape language. For Garner, language and reason entailed one another, so that evidence for the presence of language was at the same time evidence for the presence of reason. Morgan too believed that language and reason went together. Indeed, he had argued for a methodological presumption against animal reason on the grounds that animals apparently lacked language. In 1892, Morgan was well aware that Garner might return from Africa with cylinders showing that at least some animals had language after all.

The tale of Garner's phonograph thus points to an alternative career, not just for the phonograph, but for the emerging discipline of comparative psychology.[97] In *The Speech of Monkeys,* Garner offered a glimpse of the future that beckoned for scientists in literal communication, thanks to the Edison phonograph, with their own evolutionary past. Far more was in the offing than new truths about language and its origins:

> Standing on this frail bridge of speech, I see into that broad field of life and thought which lies beyond the confines of our care, and into which, through the gates that I have now unlocked, may soon be borne the sunshine of human intellect. What prophet now can foretell the relations which may yet obtain between the human race and those inferior forms which fill some place in the design, and execute some function in the economy of nature? A knowledge of their language cannot injure man, and may conduce to the good of others, because it would lessen man's selfishness, widen his mercy, and restrain his cruelty. It would not place man more remote from his divinity, nor change the state of facts which now exist. Their speech is the only gateway to their minds, and through it we must pass if we would learn their secret thoughts and measure the distance from mind to mind.[98]

Acknowledgments

Much of the research presented here was completed under the supervision of Dr. Simon Schaffer at the University of Cambridge. For help in refining ideas and prose, I am grateful to him, and also to Lindsay Gledhill, Thomas Dixon, Graeme Gooday, and the editors of this volume. For much-needed phonograph tuition, I thank Jerry Fabris of the Edison National Historic Site, West Orange, N.J., and Peter Martland.

Notes

1. Thomas A. Edison, "The Phonograph and its Future," *North American Review* 126 (June 1878): 527–536.

2. "Mr. Edison's Forecast," *Phonogram* 2 (January 1892): 1–2, quotes on p. 2.

3. "The Phonograph to Aid in Establishing the Darwinian Theory," *Phonogram* 2 (January 1892): 16.

4. On the debates over evolution in the late nineteenth and early twentieth centuries, see Peter J. Bowler, *The Eclipse of Darwinism: Anti-Darwinian Evolution Theories in the Decades around 1900* (Baltimore: Johns Hopkins University Press, 1983).

5. A better known, albeit controversial, link between Edison's phonograph and Darwin's theory is a metaphoric one. It is possible to interpret the functional winnowing of the phonograph in the later nineteenth century in terms of Darwin's theory of natural selection, such that the success of certain functions and the extinction of the others amounted to a "survival of the fittest." For full development of this interpretation, see George Basalla, *The Evolution of Technology* (Cambridge: Cambridge University Press, 1988), 135–143.

6. See, for example, Ronald L. Numbers and Leslie D. Stephens, "Darwinism and the American South: From the Early 1860s to the late 1920s," in *Darwinism Comes to America,* ed. Ronald L. Numbers (Cambridge: Harvard University Press, 1998), 58–75. For all its thoroughness, this examination of attitudes to evolution in the South of Garner's day concentrates on the writings of southern seminarians and university professor of science, and so left out Garner completely.

7. For biographical details on Garner, see the articles on Garner in *The National Cyclopaedia on American Biography,* vol. 13 (New York: James T. White, 1906), 314, and *Who was Who in America,* vol. 4 (Chicago: Marquis-Who's Who, 1968), 346. A journalist described the adult Garner as having "dark eyes, dark complexion and a very strong face. He is broad-shouldered and of great physical strength, weighing about two hundred pounds." "Can Make Monkeys Talk," *New York World* (20 December 1891), 22.

8. On antebellum Abingdon and the surrounding region, see Jack R. Garland, "An Economic Survey of Southwest Virginia during the Ante-Bellum Period" *The Historical Society of Washington County, Va. Bulletin,* ser. 2, 24 (1987): 9–15.

9. In Garner's day, for example, the present Barter Theatre, then owned by the Sons of Temperance, "was used for lectures, grange meetings, theatrical productions and . . . [from] 1855 a male school met in the basement." *Abingdon: A Self-Guided Walking Tour of Abingdon's Downtown Historic District* (Abingdon, Va.: Abingdon Convention and Visitors Bureau, n.d.).

10. Garner made two attempts at autobiography, both centering on his Abingdon boyhood. I have drawn the above mainly from an unpublished manuscript, "Just as It Happened: A True and Unvarnished Statement of Events in the Life and Experience of R. L. Garner," in the Papers of

Richard Lynch Garner, Box 4, Folder: "Writings: 'Jack-o'-Lantern Farm'–'Just as it Happened,'" National Anthropological Archives, Smithsonian Institution (hereafter "NAA/S"), Washington, D.C. On Garner's father, see p. 2; on Garner's precocious agnosticism, see pp. 3–6, quote on p. 3. Between January and May 1904, Garner wrote a series of autobiographical letters, published by his son Harry as *Autobiography of a Boy: From the Letters of Richard Lynch Garner* (Washington, D.C.: Huff Duplicating Co., 1930). On the water mill owned by Garner's father, see pp. 7–8.

11. On the wartime activities of Garner and his family, see, in addition to the biographical articles cited above, an article on Garner from around December 1916, in the *Philadelphia Evening Bulletin,* and held (in typescript) in the Urban Archives, Samuel Paley Library, Temple University, Philadelphia; and a letter from Garner's son Harry to Garner's biographer John Peabody Harrington, 22 July 1941, in the Garner Papers, Box 7, Folder: "Biographer's Papers," NAA/S. On Forrest's Raids, see Stephen Z. Starr, "Forrest's Raids," in *The Encyclopedia of Southern History,* ed. David C. Roller and Robert W. Twyman (Baton Rouge: Louisiana State University Press, 1979), 468–469. On the Virginia-Tennessee border region during the war, see V. N. Phillips, *Between the States: Bristol Tennessee/Virginia during the Civil War* (Johnson City, Tenn.: Overmountain Press, 1997).

12. After Garner visited the English newspaperman W. T. Stead in the summer of 1892, Stead wrote thus about "one of the few Americans who has recently become familiar throughout the world": "Mr. Garner is a tough customer, having served in the Confederate Army, and thus gained practical personal experience of the hardships endured by Confederate prisoners in about a dozen different northern prisons. He then spent four years in the plains campaigning against the Indians. He may therefore possibly return from Africa alive." "Off to Monkey Land: A Visit from Mr. Garner," *Review of Reviews* 6 (September 1892): 256. Thanks to Gowan Dawson of the Leeds-Sheffield SciPer project for this reference, and all subsequent references to Stead's journal, the *Review of Reviews.*

13. See Alexander Melville Bell, *Visible Speech: The Science of Universal Alphabetics; or Self-Interpreting Physiological Letters, for the Writing of All Languages in One Alphabet* (London: Simpkins, Marshall, 1867), 20–21, quote on p. 21. On "the nineteenth-century infatuation with phonetics," see Jonathan Rée, *I See a Voice: A Philosophical History of Language, Deafness and the Senses* (London: HarperCollins, 1999) 255–65, quote on p. 265. On Garner's "hard work for years as a village school-master, where he paid most attention to the study of phonetics," see Barnett Phillips, "A Record of Monkey Talk," *Harper's Weekly* 35 (26 December 1891): 1050.

14. Garner, "The Speech of Animals," unpublished ms, p. 2, in the Garner Papers, Box 6, Folder: "Writings: 'Sanctuary to Women'–'Spider Webs,'" NAA/S.

15. On the vogue of Spencer, see Richard Hofstadter's classic *Social Darwinism in American Thought,* rev. ed. (Boston: Beacon Press, 1992), chapter 2.

16. On Garner's reading, see the 1916 *Philadelphia Evening Bulletin* article cited in note 10 above, and Garner, "Critics and Croakers," unpublished ms, p. 1, in the Garner Papers, Box 3, Folder:

"Writings: 'Civilised Savagery'—'Critics and Croakers,'" NAA/S. In *The Speech of Monkeys* (London: William Heinemann, 1892), Garner wrote, of the relations between language and reason (p. 205): "I beg to be allowed to stand aside and let Prof. Max Müller and Prof. Whitney, the great giants of comparative philology, settle this question between themselves; and I shall abide by the verdict which may be finally reached."

17. See Garner, "The Simian Tongue [I]," reprinted in *The Origin of Language,* ed. R. Harris (Bristol: Thoemmes Press, 1996), 314–321, esp. 314–315. Page references will be to this reprint.

18. In "Simian Tongue [I]," Garner wrote (p. 315) that, after the Cincinnati epiphany, he pursued his translation studies without much success "in the gardens of New York, Philadelphia, Cincinnati, and Chicago, and with such specimens as I could find with the traveling menagerie museum, or hand organ, or aboard some ship, or kept as a pet in some family."

19. Garner to Prof. Brown Goode, 30 January 1888 (one letter) and 31 July 1888 (two letters), RU 189, Box 43, Folder 7, Smithsonian Institution Archives (hereafter "SIA"). On Garner's study of Maya glyphs, see also Phillips, "Record," and the contributor's note on Garner in the *Cosmopolitan* 13 (May 1892): 128.

20. Note to Prof. Brown Goode, 7 August 1888, RU 189, Box 43, Folder 7, SIA. Garner tried again the following year. Garner to Prof. Brown Goode, 2 July 1889, RU 189, Box 43, Folder 7, SIA.

21. Garner, *Speech of Monkeys,* 221–222. Garner claimed (p. 222) that, at the time of his Maya work, he was not familiar with Bell's *Visible Speech*. Garner later accused the Bureau of American Ethnology's Garrick Mallery of pirating and exploiting this method of decipherment. Garner to W. H. Holmes, 6 December 1900, in the Garner Papers, Box 1, Folder: "Outgoing Letters," NAA/S. On Mallery's earlier research into Amerindian "gesture speech," and relations between this speech and the undeciphered glyphs, see Rée, *See a Voice,* 285–292, esp. p. 285.

22. "The phonograph has made more progress in Washington, in office and home-use, than in any other city thus far." "The Columbia Phonograph Co., Washington, D.C.," *Phonogram* 1 (April 1891): 88–92, quote on p. 89.

23. Phillips, "Record."

24. On Edison's invention of the phonograph, and the importance of the phonograph in turning Edison into "the wizard of Menlo Park," see Paul Israel, *Edison: A Life of Invention* (New York: John Wiley and Sons, 1998), 142–156. On forerunners of the phonograph, see Oliver Read and Walter L. Welch, *From Tin Foil to Stereo: Evolution of the Phonograph,* 2d ed. (Indianapolis: Howard Sams, 1976), chapter 1.

25. On the rapid rise and fall of the tin-foil phonograph, see Roland Gelatt, *The Fabulous Phonograph: 1877–1977,* 2d rev. ed. (London: Cassell, 1977), chapter 1, and Read and Welch, *Tin Foil,* 21–26. For a lucid explanation of how early "acoustic" phonographs work, see Erika Brady et al., *The Federal Cylinder Project: A Guide to Field Cylinder Collections in Federal Agencies* vol. 1 (Washington, D.C.: American Folklore Center/Library of Congress, 1984), 4–6.

26. Israel, *Edison,* chapters 10–12, and Read and Welch, *Tin Foil,* 26.

27. On Edison's perfecting of the phonograph, see Israel, *Edison,* 277–291, Andre Millard, *Edison and the Business of Innovation* (London: Johns Hopkins University Press, 1990), chapter 4, and Gelatt, *Phonograph,* chapter 2.

28. "A Retrospect," *Phonogram* 1 (September 1891): 187.

29. The year 1891 was the centenary of Faraday's birth. For a discussion of this cartoon in the context of the centenary more generally, see Graeme Gooday, "Faraday Reinvented: Moral Imagery and Institutional Icons in Victorian Electrical Engineering," in *History of Technology* vol. 15, ed. Graham Hollister-Short and Frank A. J. L. James (London: Mansell, 1993), 190–205, esp. 193–194.

30. On Edison's visit to the Paris Exposition of 1889, where the perfected phonograph was first exhibited on the Continent, becoming "the most notable and popular device at the exhibition—every day, some 30,000 people heard twenty-five phonographs talking in dozens of languages," see Israel, *Edison* 369–371, quote on p. 371.

31. Gelatt, *Phonograph,* 43–44.

32. Israel, *Edison,* 287–290 and Gelatt, *Phonograph,* 45.

33. Israel, *Edison,* 291, and Gelatt, *Phonograph,* 43–44.

34. "Columbia Phonograph Co.," 89–90.

35. On the nickel-in-the-slot phonograph, see Charles Musser and Carol Nelson, *High-Class Moving Pictures: Lyman H. Howe and the Forgotten Era of Traveling Exhibition, 1880–1920* (Oxford: Oxford University Press, 1991), 38–40 and Gelatt, *Phonograph,* 44–57. "It is, perhaps, not altogether conducive to the higher aims and nobler uses of the phonograph to cause its initial exploitation as a catch-penny adjunct of the saloon-keeper," wrote one *Phonogram* author, "but, on the other hand, it brings it, in these public places, before the public eye, where its simplicity and wonder-creating functions will quicken the public pulse to a superficial contemplation calculated to develop a serious consideration of its possibilities in the world of affairs." William Addison Clark, "Phonographic Possibilities," *Phonogram* 1 (April 1891): 86–87, quote on 86.

36. On the Bureau of American Ethnology, see Curtis M. Hinsley, *Savages and Scientists: The Smithsonian Institution and the Development of American Anthropology, 1846–1910* (Washington, D.C.: Smithsonian Institution Press, 1981).

37. Jesse Walter Fewkes, "On the Use of the Phonograph in the Study of the Languages of American Indians" *Science* 15 (1890): 267–269, quotes on 267, 268 and 269. On the phonograph among anthropologists more generally, see Erika Brady, *A Spiral Way: How the Phonograph Changed Ethnography* (Jackson: University Press of Mississippi, 1999).

38. Garner's experiments had also involved Bell's graphophone rather than Edison's phonograph (though the term "phonograph" was commonly used for both). See "Can Monkeys Talk? A Queer

Scientific Experiment at the Smithsonian Institute," *St. Louis Daily Globe-Democrat* (21 September 1890). Two months later, Garner spoke again to a reporter from the same newspaper about new experiments in Chicago and Cincinnati. "Monkey Talk of the Future," *St. Louis Daily Globe-Democrat* (22 November 1890). Thanks to Pamela Henson of the Smithsonian Institution Archives for these references.

39. Frank Baker to Garner, 8 December 1890, RU 74, Box 7, Folder 2, SIA. On the National Zoological Park and Baker's role there, see "National Zoological Park," *Guide to the Smithsonian Archives 1996* (Washington, D.C.: Smithsonian Institution Press, 1996), 187–188, esp. 187.

40. "Novel Experiments with the Phonograph," *Phonogram* 1 (April 1891): 85.

41. Garner, "Simian Tongue [I]," 314. The article was published in England in the *New Review* 4 (June 1891): 555–562, and in the United States in *Littel's Living Age* 190 (1891), 218–221. Abridged versions appeared in the *Review of Reviews* 3 (June 1891): 574 and the *Phonogram* 1 (September 1891): 200–202. For an analysis of Garner's argument for the simian tongue as ancestral to human languages, see G. Radick, "Morgan's Canon, Garner's Phonograph, and the Evolutionary Origins of Language and Reason," *British Journal for the History of Science* 33 (2000): 3–23, esp. 13–14.

42. "Simian Talk," *Punch* 100 (20 June 1891): 289, emphases in original. Much the same joke appears in the notebooks of the anti-Darwinian man of letters Samuel Butler for the next year: "In his latest article (February 1892) Prof. Garner says that the chatter of monkeys is not meaningless, but that they are conveying ideas to one another. This seems to me hazardous. The monkeys might with equal justice conclude that in our magazine articles, or literary and artistic criticisms, we are not chattering idly but are conveying ideas to one another." See *The Note-books of Samuel Butler,* ed. Henry Festing Jones (London: A. C. Fifield, 1912), 185. Thanks to John van Wyhe for this reference.

43. William James, *The Principles of Psychology* in the Harvard edition of 3 volumes, comprising the 2 volumes of the 1890 original (see vol. 2, p. 980, for the passage cited above) and an additional volume of James's annotations etc. (see vol. 3, p. 1470, for the annotation cited above). *The Works of William James,* vols. 8–10 (London: Harvard University Press, 1981).

44. "Apes and Men," *Spectator* 66 (6 June 1891): 787–788, quote on 788.

45. For an overview of science in British periodicals during the later nineteenth century, see William H. Brock, "Science," in *Victorian Periodicals and Victorian Society,* ed. J. D. Vann and R. T. VanArsdel (Aldershot: Scolar Press, 1994), 81–96. For an overview of science in American periodicals during the same era, see Frank L. Mott, *A History of American Magazines,* vol. 4 (Cambridge: Belknap Press/Harvard University Press, 1957), 306–310.

46. On the *New Review,* see *The Wellesley Index to Victorian Periodicals, 1824–1900,* ed. W. Houghton and E. R. Houghton (London: Routledge/Kegan Paul, 1979), 303–308.

47. On McClure, see R. L. Gale's biographical article in *American National Biography,* vol. 14, ed. J. A. Garraty and M. C. Canes (Oxford: Oxford University Press, 1999), 887–889. On Mc-

Clure's connections in English publishing, see esp. H. S. Wilson, *McClure's Magazine and the Muckrakers* (Princeton: Princeton University Press, 1970), 48–55. On McClure's placing of Garner's works with Grove's *New Review,* see Garner's remarks on his 1893 *McClure's Magazine* article, in Garner, "Index to Manuscripts with Notes on Publications, Transmission, etc.," in the Garner Papers, Box 2, Folder: "Writings: Indexes," NAA/S.

48. The earliest evidence I have found of the relationship between Garner and McClure is a letter from late 1891. Garner to S. S. McClure, 24 November 1891, in Collection no. 9040, the Alderman Library, University of Virginia. In his autobiography, McClure remarked on how rare were the interesting scientific experts, and how valuable to publishers for the general public. Samuel Sidney McClure, *My Autobiography* (London: John Murray, 1914), 244.

49. See various letters and announcements in the Letter Copybooks from 1891–1892 and 1892–1896, in the Papers of Samuel Sidney McClure, Lilly Library, Indiana University, Bloomington, Indiana. Thanks to Dina Kellams for locating these documents.

50. "Can Make Monkeys Talk," *New York World* (20 December 1891), 22.

51. Phillips, "Record."

52. "Speech in the Lower Animals," *Scientific American* 66 (27 February 1892): 129.

53. Garner, "The Simian Tongue II" and "The Simian Tongue III," in Harris, *Origin* 321–327 and 327–332, respectively. Both appeared originally in the *New Review* 5 (November 1891): 424–430 and 6 (February 1892): 181–186. Garner was paid $65 for the first installment of "The Simian Tongue," $35 for the second, and $65 for the third. Garner, "Index."

54. Garner, "Simian Speech and Simian Thought," *Cosmopolitan* 13 (May 1892): 72–79, and "The Speech of Monkeys," *Forum* 13 (April 1892): 246–256.

55. On the origin of and profits from *The Speech of Monkeys,* see Garner, "Index." Previously Garner had privately published two books of verse. In *Nancy Bet: The Story of Sloomy Perkins and His Transaction in Real Estate* (1889), the narrator tells of having married ugly Nancy forty years before, in hope that her ailing father would pass away and leave them his farm. Alas, the old man improved, and Nancy turned out to be an insufferable shrew. After much mayhem ("She'd banged my eyes till they went black, / And poured hot ashes down my back . . ."), they divorced. The poems collected in *The Psychoscope* (1891) are far more serious and contemplative, with somber titles such as "Perfidy," "Despair," and "The Church Militant."

56. Garner, *Speech,* 43.

57. Garner, *Speech,* 103.

58. Garner, *Speech,* 3.

59. Reviews appeared in, among other publications, *Nature, Science, Popular Science Monthly,* the *American Anthropologist,* the *American Naturalist,* the *Spectator,* the *Dial,* the *Critic,* the *Nation,* the *Athenaeum,* the London *Times,* the *New York Times,* the *New-York Daily Tribune,* the *Brooklyn Daily*

Eagle and the Dutch journal *De Gids*. The Russian edition, *Yazyk obez'yan,* came out in 1899, and the German edition, *Die Sprache der Affen,* in 1900. A second German edition followed in 1905.

60. *Spectator* 69 (13 August 1892): 227–228, quote on 228. In all likelihood, the *Spectator* reviewer was C. J. Cornish, as the review appears verbatim in the chapter on "The Speech of Monkeys" in Cornish's 1895 book on Regent's Park Zoo, *Life at the Zoo: Notes and Traditions of the Regent's Park Zoo* (London: Seeley, 1895). Thanks to Sonia Åkerberg for this reference.

61. Joseph Jastrow, "Conversations with the Simians," *Dial* 13 (1 October 1892): 215–256. Like the more positive *Critic* reviewer, however (see the *Critic* 21 [24 September 1892]: 161–162), Jastrow (p. 215) described some nonphonographic experiments testing "general intelligence" as "the most valuable in the book. The method of testing the counting powers of monkeys is ingenious, and some of the tales illustrating their successes in adapting means to ends form welcome contributions to our stock of observations." The Dutch reviewer A. A. W. Hubrecht, at least as translated in a brief excerpt in the *Review of Reviews,* took an unrelievedly dim view of Garner and his project: "When Garner, in his semi-naive communications, not only sells the skin before catching the bear, but tries to coin money out of the various apparatus with which the capture is to be effected, we see in this a new proof of his want of that seriousness and critical sense which is absolutely indispensable to the success of any scientific investigation." "Spreken De Apen?" *De Gids* (December 1892), 508–522, abridged in translation in the *Review of Reviews* 7 (January 1893): 61.

62. E. P. Evans, "Studies of Animal Speech," *Popular Science Monthly* 43 (August 1893): 433–449, quote on 438.

63. Evans, "Studies," 438.

64. Evans, *Evolutional Ethics and Animal Psychology* (New York: D. Appleton, 1897), 315.

65. Phillips, "Record."

66. Garner, "A Monkey's Academy in Africa," *New Review* 7 (1892): 282–292, quote on 289.

67. Garner, "Monkey's Academy," 290.

68. "Professor Garner has the special patronage of ex-President Grover Cleveland and Mr T. A. Edison in his unique undertaking." *Yorkshire Post,* mid-September 1892, quoted in E. H. T., "Revolution" *Blackwood's Edinburgh Magazine* 153 (February 1893): 253–255, on 253.

69. Garner to T. A. Edison, 27 July 1891 and 25 August 1891, in the Papers of Thomas A. Edison, microfilm/online ed., reel 131, frames 851–853 and 857–858, respectively, quote on Frame 853. (Hereafter cited as, e.g., "TAEM 131: 853.")

70. Edison's notes for George Lathrop, n.d., TAEM 128: 749. On *Progress*, see Israel, *Edison,* 365–368. Thanks to Paul Israel for this reference.

71. A. O. Tate to Garner, 31 August 1891, TAEM 142: 758–759, quote on frame 759.

72. Garner to Edison, 21 December 1891 (with Edison annotation dated 24 December 1891), TAEM 131: 868–870. In the letter Garner mentioned that it was "Prof. Bell" himself who had advised Garner on switching from batteries to springs.

73. "Mr. Garner Talks with Monkeys" *New York Herald* (10 December 1891): 14.

74. See, e.g., Israel, *Edison,* 286.

75. See, e.g., Israel, *Edison,* 132–141, 234–237.

76. "Mr. Edison's Forecast," quote on p. 2.

77. "Speech in the Lower Animals."

78. Garner, "A Monkey's Academy," 285–286.

79. G. N. Morison to Edison (with Edison annotation), 1 April 1892, TAEM 133: 329.

80. Garner, *Speech,* 208–209.

81. The following analysis owes much to Ian Hacking's writings on experiment. See esp. "The Self-Vindication of Laboratory Science," in *Science as Practice and Culture,* ed. Andrew Pickering (Chicago: University of Chicago Press, 1992), 29–63, "Artificial Phenomena [review of S. Shapin and S. Schaffer, *Leviathan and the Air Pump*]" *British Journal for the History of Science* 24 (1990): 235–241, and *Representing and Intervening: Introductory Topics in the Philosophy of Natural Science* (Cambridge: Cambridge University Press, 1983), chapter 13.

82. Garner, *Speech,* 221.

83. Garner, *Speech,* 211.

84. Garner described the procedure in *Speech,* 210–212.

85. Garner, *Speech,* 73.

86. Garner, *Speech,* 111.

87. Cf. Garner, *Speech,* 201: monkeys showed they understood the meanings of monkey utterances even when those utterances were "imitated by a human being, by a whistle, a phonograph, or other mechanical devices, and this indicates that they are guided by the sounds alone, and not by any signs, gestures, or psychic influence."

88. Garner, *Speech,* 61–62.

89. Garner, *Speech,* 210.

90. Garner, *Speech,* 209–210.

91. For other instances of phenomena creation, see Garner, *Speech,* 181–182, 209, 212, 215–216, 219–121.

92. Garner, *Speech,* 211–212. Garner continued to explain what he took to be unorthodox views on the nature of sound waves on pp. 212–214.

93. Garner, *Speech,* 213. On the phoneidoscope, see Sedley Taylor, "Sound Colour-Figures," *Nature* 17 (28 March 1878): 426–427, which identifies the manufacturer of the phoneidoscope as S. C. Tinsley and Co., Philosophical Instrument Makers, London. A phoneidoscope is held in the Whipple Museum, Cambridge.

94. Frank Luther Mott, "The Magazine Revolution and Popular Ideas in the Nineties," *Proceedings of the American Antiquarian Society* 64 (1954): 195–214, quote on 205.

95. For the story of how Garner and his phonographic achievements came to grief after his expedition, due in part to the conditions of the "New Journalism," see Gregory Radick, *Animal Language in the Victorian Evolutionary Debates,* chapter 4 (Ph.D. dissertation, Cambridge University, 2000).

96. For a superb discussion of *Truth*'s combination of society journalism and what we now call investigative journalism, see G. Weber, "Henry Labouchere, *Truth* and the New Journalism of Late Victorian Britain" *Victorian Periodicals Review* 26 (1993): 36–43.

97. I develop this argument more fully in Radick, "Morgan's Canon."

98. Garner, *Speech,* 110.

9 Scissorizing and Scrapbooks: Nineteenth-Century Reading, Remaking, and Recirculating

Ellen Gruber Garvey

Everyone who takes a newspaper . . . will often regret to have it torn up, on account of some little scrap in it which was of importance to them. . . . If you will hoard these rare gems, year after year, you will garner up a treasure-book that will not only be of service to yourself, but also to your children and grandchildren, in decades of years yet to come.[1]

In a 1914 memoir of a Missouri valley childhood, Margaret Lynn recalled her brothers and sister making scrapbooks together. The children took old bound volumes of *Reports* of proceedings of agricultural and horticultural societies and pasted over them: "We seemed to feel it laid on us to preserve the literature of newspapers and magazines from utter oblivion by entombing it in the sarcophagus of a scrapbook. I think each of us had at least one to his credit annually." The children vied for desirable printed matter to use in their books. "A hazy aunt somewhere subscribed for *Our Young Soldiers* for John," writes Lynn, and "[e]xcept for the pleasure of having it come addressed to him, he did not care much for it, and generously allowed us to read it freely. . . . But on these occasions he claimed everything usable in it, and only flung us the hacked and rifled remains." Choosing between articles on opposite sides of the same sheet posed difficulties: "Which should be saved for immortality and which should be lost forever in the pasty act of adhesion?"[2]

Michel de Certeau's metaphor *poaching* to describe readers' appropriations of texts to their own uses and meanings would seem a particularly apt description of the Lynn children at work. De Certeau distinguishes readers from writers, who are "founders of their own place . . . diggers of wells and builders of houses." Readers instead "move across lands belonging to someone else, like nomads poaching their way across fields they did not write, despoiling the wealth of Egypt to enjoy it themselves."[3] De Certeau's poaching is an act of despoiling, in which a reader runs over one text to place his or her own ideas in it, yet after de Certeau's reader is done with the page, the text remains intact, available to its next poachers. His metaphor for reading is more literally descriptive of scrapbook making: The scrapbook maker actually despoils—clipping a paper full of holes, and as in the case Lynn describes, pasting over another work.

In his work on *Star Trek* fans, Henry Jenkins has corrected de Certeau's conception, pointing to its mistaken teleology, its implication of a "correct" reading that poachers violate, as well as to its neglect of community building potential, as when poachers' collective activity integrates them within a subcultural community.[4] Jenkins's version of "textual poaching" comes closer to describing the act of creating new media from old at which scrapbook makers industriously plied their scissors. Yet the metaphor that scrapbook makers themselves more often used for their work was not *poaching* but *gleaning*: gathering the small bits, the leftovers, the dropped grapes or the grain left in the corners of the field—the surplus and excess—and making a meal of them. (One scrapbook seen for this study even has a print of Jean Francois Millet's painting "The Gleaners" on its cover.) Like the poacher, the gleaner does not own the land, did not produce the crop or livestock, but steps in when it is ready, takes what is available, and puts it to her or his own uses. Gleaning shifts from the implied masculinity of shooting game, engaged in a kind of warfare with the landowner, to a model of gathering that is not passive or compliant, and is decidedly open to feminine participation. Poaching has important reciprocal elements: The gleaners' bounty depends on the planter's willingness not to squeeze every possible bit of profit from the land; following the biblical injunction to leave food for the gleaners, the planter does not pursue every last scrap of grain or produce but leaves some unclaimed.[5] The gleaner can still create multiple meanings and readings from the text, and can even bake bread from gleaned grain and sell it under the gleaner's label.

As literary critics have long noted, authors inevitably leave a surplus of meaning, sometimes obvious as ambiguity, which readers maneuver within, or scoop up, glean, and reuse. And just as authors cannot nail meaning to a fixed spot, neither can they or their publishers control the circulation and ordering or reordering of meaning. Even when copyright locks down the right to reproduce texts, readers have the option of moving those old texts to new contexts, creating a new tier of private circulation: clipping texts out of newspapers, pasting them into scrapbooks, or today onto Web pages, and circulating this new compiled version. Nineteenth-century scrapbook makers were part of an elaborate circuit of recirculation, one that trespassed or found easements across the enclosure of authorship and publication.

Coping with Old Media

Many nineteenth-century American readers found books expensive or hard to obtain, while newspapers and magazines proliferated at an overwhelming rate. "We have so many old newspapers that we cannot afford house-room for them all; and if we could, they would never get re-read. Old papers rarely do," Julia Colman complained in 1873.[6] They constituted a new subcategory of media—the cheap, the disposable, and yet somehow tantalizingly valuable, if only their value could be separated from their ephemerality. Readers adapted to this proliferation of print by cutting it up and saving it, reorganizing it, and sometimes recirculating it. Approaches for coping with the rising tide of print included library vertical files and cataloging systems like Melvil Dewey's, and commercial clipping bureaus that scanned a city's or nation's papers on behalf of clients. But at home, other readers, like a family Colman visited, created scrapbooks to capture this value. The clipping scrapbooks I discuss here were generally a separate category from memorabilia scrapbooks and related works like theatrical scrapbooks that we associate with documenting personal lives. But each clipping scrapbook maker too created a private and idiosyncratic catalog, a reflection of personal identity made from mass-produced and distributed publication as much like Netscape Bookmarks or Microsoft Favorites as a library's vertical files. These Web management devices help users blaze a trail through a vast landscape of materials.

A variety of technologies has long existed to assist readers in managing the confusing abundance of texts. Three-dimensional bookmarks allow the reader to find the same place again in a codex, making it easier to compare passages in different parts of the same book, or in different books. Medieval bookmarks that consisted of a bead or other anchor with a bundle of ribbons or threads running out from it, marking many pages in the same book at once, eased the comparing of passages. Accessing several passages at a time was further facilitated by the Renaissance bookwheel, a specialized apparatus holding many open books. The seated reader turned it to move between desired passages, holding them all open for reference, presumably while distilling them into a new form, writing something else from them.

The bookwheel assumed a limited number of books would be wanted at one time; the three-dimensional bookmark assumes that the reader will recall what book the desired passage was in. Another system, John Todd's *Index Rerum* of 1835, encouraged readers to create a personal catalog of their own reading. Todd proposed a system for tying one's reading to an alphabetized catalog, so that invisible threads run out from the index to bookshelves. The scheme was offered in the form of a blank book with pages headed alphabetically, with each letter subdivided by vowels (it begins Aa, Ae, Ai, etc.); the user entered a word on the page matching the first letter and first vowel following, noting an item pertaining to "alphabet," for example, on the page Aa.

The *Index Rerum* assumed continued access to the books one referenced, in stable form: the same old editions, so that the page numbering would hold true; a unified, organically growing self, so that the lines that struck the reader at age twenty would be the passage still needed at age forty. It assumed that items would be categorized well enough to be found again; and of course that the reader would be organized enough to keep the *Index Rerum* at hand, and be diligent enough to take the notes in the first place.

The commonplace book, in which passages from other works are copied out by hand, in use for centuries, serves some of the functions of both the bookwheel and *Index Rerum*, but instead of marking a place in the cornfield, it scoops up the grain and turns it into a new form, bakes bread with it, makes a new book of it. A trail of crumbs or references may mark the way back, but more likely leaves the reader of the commonplace book with something new: a collection of passages that lead nowhere, but have become freestanding "quotes" and sayings. The reader becomes an author. In her memoirs, Kate Sanborn, a late nineteenth-century author, lecturer, and anthologist, cites a verse on the subject she learned as a child:

> In reading authors, when you find
> Bright passages that strike your mind,
> And which, perhaps you may have reason
> To think on in another season;
> Be not contented with the sight,
> But jot them down in black and white;
> Such respect is wisely shown
> As makes another's thought your own.[7]

She memorized the poem—another way of making another's thought one's own. And while memorizing it and copying it, Sanborn dropped the author's name, very effectively making it her own. Depending on how well the reader notes the sources, the trail of crumbs left behind may leave the reader stranded without a way back, like Hansel and Gretel after dropping their bread crumbs in the woods.

As John Todd noted in presenting his *Index Rerum,* commonplace book keeping required more effort than most readers would put into it; it was not well adapted to coping with the increased quantities of materials passing before readers' eyes. Even following Todd's paradoxical injunction to users of his volume to "Read nothing which is not worth remembering and which you may not wish hereafter to review" wouldn't tame the wilderness to be tracked.[8] The *Index Rerum* is tied to an ideal world of permanence: books that stay on a predictable shelf, or at least maintain the same form and pagination. The books' permanence invites respect: The reader should not mark them or cut out useful portions, but preserve their integrity, and hope that the threads of the *Index Rerum* leading to them would be strong.

By the final quarter of the nineteenth century, the distance between the ideal books Todd instructs readers to index and the real reading material they encountered widened. Some readers were likely to read many more books than they would be likely to own, leaving them with the reference to a book they could no longer find, or a page reference that did not match an edition they now had access to—an expired link, in other words. More significantly, magazines and newspapers proliferated, and these were not as likely to remain on a shelf or offer continued access. An 1892 *Harper's Monthly* column saw the shift as symptomatic of a more general problem of the accelerating pace of life: "The magazine, in a generation that must run as it reads, takes the place of the book," the columnist complained. "Must we all go to making scrap-books in order to preserve the good things that fly on the leaves of the winged press?"[9]

The virtual and three-dimensional bookmark, the *Index Rerum,* the commonplace book, and the scrapbook all have in common the desire to mark the path of one's own reading, to *not* pass through the field without leaving a trace, but to find one's way back to a desired spot. That path might be a public one, in the form of the links on a widely advertised Web page, or a published *Index Rerum* or selection of favorite sayings; it might be privately kept as a scrapbook on a hidden shelf, a commonplace book in code, or on a password-protected home computer; or it might be somewhere between these extremes: a scrapbook on the parlor table, borrowed by friends; or a personal Web page unlikely to show up on a search because few or no other sites are linked to it.

The scrapbook had some advantages over the *Index Rerum,* but it possessed a different set of flaws. It emphasized its own inadequacy; it could not encompass all reading that interested a reader; it could not hold even every poem one might wish to read again, and if it did, it would be too big to be useful, returning the reader to the original problem of surplus. Even focused or classified scrapbooks are miscellaneous; they may have started out to hold or categorize one kind of material, but materials ultimately slip out of category, demonstrating their instability. In reusing both the form of the book and the clippings pasted in, scrapbooks resemble websites, which mingle the link-compiling function of the bookmark with content borrowed without attribution from other websites. The same recycled source codes underlie thousands of websites, just as scrapbooks silently borrowed their formal qualities from one another and from the volumes that structured them.

Although it is easy to see the advantages of capacious, expandable, fluid hypertext over three-dimensional scrapbooks constrained by paper and glue, scrapbooks have at least one significant advantage: Links in scrapbooks do not expire. Scrapbooks thus both embody a system of categorizing material, however haphazard and full of exceptions, and leave a permanent record of that categorizing. Because they literally—materially—categorize, scrapbooks can reveal something of how readers read and thought of the works they saved; because they mirrored and brought into the home the practices of recirculation that were common in the press, they can reveal something of nineteenth-century American attitudes about the boundaries between reading and authoring and about the ways in which all media, old or new, are experienced as renewable resources.

Exchanges

While readers clipped papers and made scrapbooks at home, in the professional realm of publishing, editors engaged in the related practice of exchanging. Nineteenth-century publications routinely picked up and reused material from other publications without payment and sometimes without credit. *Exchange* might suggest a simple one-for-one passing back and forth of papers, but as we might guess from some of its other uses, such as the telephone exchange and the stock exchange, the term also comprises circuits of connection and diffusion. The value of exchanges as a method for spreading news and information around the country was recognized by the post office, which allowed newspapers to send their exchange papers free. Some editors were said to send out from "ten to sixteen hundred exchange papers to other editors across the country," and more internationally.[10]

The exchanges among publications allowed newspapers in one locality to pick up and run news from another and helped to create a local press that covered the nation without specific correspondents in each locality. Journalists and commentators often remarked on this recirculation of material, not as a failure to generate original material, but as a virtue—a mechanism unifying the country. As one journalist asserted, "A man who reads the daily exchanges of the country may see an idea travel from the Atlantic slope to the Pacific and from the Pacific to the Atlantic as visibly as a train of freight cars runs over the Vanderbilt system."[11]

Going through other papers for exchanges was a routine part of a periodical staff's work. Editorial shears were considered as indispensable as the editor's pencil or printer's composing stick. Large papers had a special "exchange editor" with this duty. On smaller publications it was part of the general editor's work. So, for example, as a typical small town editor, Bartley Hubbard, of William Dean Howells's *A Modern Instance,* makes up the week's paper by going through other newspapers with his editorial scissors.

The exchange mechanism's potential for proliferation was attractive. One missionary writer suggested hopefully that if even one column "of interesting religious matter could be introduced into each of [the nation's many] papers, it would be equivalent to the annual distribution of more than sixteen hundred million tract pages."[12] But casting a column upon the waters was no guarantee of seeing it reprinted a thousandfold. The exchange process was a selective one, because editors engaged in a gatekeeping or winnowing process. In the case of signed columns, exchanges were therefore understood as a spontaneous, decentralized index to the popularity of a writers' work. So in Fanny Fern's *Ruth Hall* the indication that Ruth is achieving acclaim as a writer is in the number of her pieces that are "scissorized" and copied into exchanges: "A good sign for you Mrs. Hall; a good test of your popularity," her publisher tells her. And her brother, the spoiler, editor of another periodical, instructs his staff not to copy his sister's work, so as not to give her further publicity.[13] In Lillie Devereux Blake's 1874 novel *Fettered for Life,* similarly, Frank Haywood, reporter for the *Trumpeter,* is succeeding when his reports on his travels in the South "meet with the highest praise from all quarters, and are copied all over the country."[14] This free access allowed to editorial gleaners could enhance the publisher's as well the author's stature. In editor Jeannette Gilder's novel *Taken by Siege* (1886), a young journalist's "local stories began to be largely copied by the State papers, and *The Freelance* got a reputation that it had never had before."[15]

Many publications carried a column headed "from our exchanges" marking the material as coming from outside the community. Exchange columns spread ideas and perhaps created a sense of identity among such groups as readers of high school newspapers, fraternity members, and suffragists, all of whose newspapers carried reports from kindred newspapers via exchanges.

Clipping and Making Scrapbooks

While editors were plying the editorial shears in the office, a parallel activity was under way in the home. (I have elsewhere discussed the gendered element to these activities.[16]) Readers treasured and saved up stories and articles in scrapbooks, creating permanent repositories that reflected their tastes and uses for works. Saving and reusing scraps was an ideal of housekeeping. Just as cloth scraps were to be saved and recycled as new clothes or as quilts, scraps of usable printed matter were saved and remade in another form. The two forms were occasionally compared. Like quilters, scrapbook makers generated a new form along the way, *creating* the usefulness of the scraps by saving and classifying, and thus making them available for reuse. The scrapbook was often understood as a liminal form between book and newspaper, and making scrapbooks was described in nineteenth-century books and articles on the subject as taking place on the border between reading and authoring.

Like newspaper exchanges, the contents of scrapbooks were understood as an index to the popular heart: Critics complained that deserving American poets had never received book publication and therefore had "no habitation but the corner of a newspaper, or the scrapbook of a friend."[17] Admiration for a poem was expressed by saying that it would be clipped for many scrapbooks, while writer-characters within several stories and poems find having their work in someone's scrapbook the highest tribute. The scrapbook asserted nondominant, if not subversive, readings. In fact the acclaim of the scrapbook was sometimes set against the judgment of more prestigious authorities: "If the scrapbooks of the land could to-day be drawn forth from their receptacles, we should find that Alice Cary has a place as a poet in the hearts of the people, which no mere critic in his grandeur has ever allowed," says one writer in a typical version of this move.[18] But unlike Henry Jenkins's *Star Trek* enthusiasts, all those Alice Cary collectors were not likely to gather or in other ways form a community over their scrapbook making.

Like exchanges, scrapbooks endorse an ideal not of originality, but of reuse and recirculation, of making the old continually new. In fact, they rotated in the same circuit with exchanges. E. W. Gurley, writing on scrapbook making, advises, "As a general thing, original articles are not the best. Those items, stories, and poems which have gone the rounds of the press until they have had the corners knocked off, have proved their worth by their circulation; it is 'the survival of the fittest.'"[19] A similar Darwinian emphasis appears in another nineteenth-century scheme for classifying and making excerpts useful, *John Bartlett's Familiar Quotations,* first issued in Boston in 1855, which valued the long-lived *familiar* quotation over the surprising or little-known one. Gurley goes on to find further virtues in widely circulated work: "The best papers to select from are those with 'patent insides'"—that is, papers with syndicated ready-printed material sent out to papers across the country. "Extracts from such papers are usually well printed, and the extracts from them look well in the scrap-book."

One could participate in literary culture, in other words, by joining in this work of recirculation, or even by making attractive books. In the most extreme version of this activity, re-creating the form of a book might override the possibility of reading it. Margaret Lynn and her siblings, all bookish children, were sometimes particularly attentive to this aspect of scrapbook making: "the nice management which brought everything out even at the bottom of the page." Her brother Henry, who didn't like poetry, was eager to have it for his scrapbook, to make attractively even columns: "Mary, with her customary readiness of device, filled in her inch or half-inch spaces with miscellaneous obituary notices. It didn't matter if she didn't know the people, she said; they were dead just the same." John trimmed a poem "to fit a vacant space, impartially clipping off the last words on the outstanding ends of the longer lines."[20]

Making Books

Makers of clipping scrapbooks used a variety of forms for their collections. Some used the elaborately bound books sold as scrapbooks and more often used for trade cards and calling cards. Some booklike systems sold commercially were intended for quasi-professional use by speakers. Mark Twain marketed his patent preglued scrapbook, suggesting it would forestall the utterances of profanity that resulted from not finding the gluepot when wanted. Margaret Lynn found that the gift of a Twain scrapbook obligated her to "choose carefully and economically the matter that was to be perpetuated in its orderly pages." It "almost put the making of scrapbooks on a new plane, and forbade the use of inferior material."[21] The fact that it cost something hindered her former freewheeling scrapbook-making ways.

Nonprofessional readers were much drawn to the form of the book, which conferred dignity and authority on the clippings. Blank books bestowed permanence on clippings and housed fugitive pieces. And many scrapbook makers reused old books. Lynn and her siblings used Agricultural and Horticultural Reports, a box of which were delivered to their home each year. Their tedious contents made them more empty than blank books, and making them into scrapbooks redeemed them from limbo.

Figure 9.1   Scrapbook Systems, advertisements in *Overland Monthly* 17, no. 102 (June 1891): xxiv.

From our point of view the books were quite unreadable and almost pitifully useless. A book that couldn't be read was an abject thing. . . . It didn't seem possible that so many books should be published with absolutely nothing in them.[22]

Julia Colman's 1873 article on scrapbook making advocated this practice. The narrator, visiting the scrapbook-making family we met earlier, finds the mother and children busy:

> Each child had a brush, a pair of scissors, some waste paper, and a book. In the latter [Hubert] was neatly pasting the slips which he received from the mother."

The narrator is shocked.

> "Why,' said I. 'You are using up good printed books!"
>   "Good for what!" was the Yankee reply, more easily made than gainsaid. "There is nothing in them that we want, and so we propose putting in something, rather than have them stand idle. Hubert's, you see, is an old day-book, and we have one or two others. Some of them are old school-books, not much worn, but out of date. Almost every library has some useless books."[23]

The mother explains that buying blank scrapbooks would be too expensive, since each child has at least two scrapbooks to accommodate his or her various interests. The editor of the California magazine *Overland Monthly* describes in 1896 finding the scrapbook he kept as a young man and wrote, "It is an old 'Agricultural Report,' and emits a damp, aged

MARK TWAIN'S PATENT SCRAP-BOOK

Gummed, ready to receive
your Scraps.

No paste or mucilage required.

Prices from $1.25 to $3.50
each including postage.

Send for descriptive Circular.

SLOTE, WOODMAN & CO., 119 & 121 William St., N. Y.

Figure 9.2   Advertisement for Mark Twain scrapbook, *Scribner's,* May (?)
1877, 11.

odor."[23] He assumes his readers will understand this compression, without stopping to explain that he made his scrapbook by pasting clippings into an old book.

Reused books commonly included used ledger books and day books, which might obviously have a limited utility once they'd been used, but also novels like volume I of *Thaddeus of Warsaw,* or a religious text like *The Saint's Everlasting Rest* by Oliver Cromwell's chaplain Richard Baxter, and even Bibles. Reusing books to make scrapbooks goes beyond Yankee thrift. A clipping in the Baxter scrapbook, which seems at first perverse in the context of a book that has pieces of paper glued over its leaves, and other leaves cut out to make room for the clippings, points to one meaning of the choice of an old book for a scrapbook. The unattributed domestic advice piece, "[ ]s for Using Books" (the first word is illegible), is a list of rules for the physical care of books: "Never hold a book near a fire. Never drop a book upon the floor. . . . Never turn down corners of leaves. . . . Never write upon a paper laid upon the leaves of an open book, as the pencil or pen point will either scratch or cut the book leaves."[25] The clipping hints that despite having in one sense destroyed a book, the maker valued books—prized them as a symbol of qualities worth preserving and saw the binding and the rest of the physical book as valuable. The value she gave her ephemeral clippings by pasting them in it placed the clippings with books worthy of protection. Even her slicing out of leaves in between pages with clippings has the effect of preserving the binding by not overfilling the book. So this scrapbook maker's claim to authorship through making a scrapbook was further buttressed by making her scrapbook as booklike as possible.

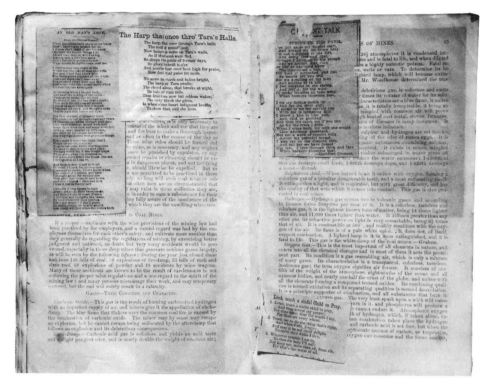

Figure 9.3A–D   Books reused as scrapbooks, collection of the author. A. and B. *Reports of the Inspector of Mines,* 1875. Publication information obscured. C. Ledger book. D. *Senate Documents: Annual Report of the Commissioner of Patents.* Publication information obscured. Inscribed: "N.B. This Book was made by Erskine C. Grice, Franklin, [?] 1872. Read it with care."

Scrapbooks were a domestic form, like the "profiles" on a personal computer, storing the user's choices for what should be on the desktop, for example, yet the circuit of recirculation within which they were engaged was neither wholly domestic nor wholly public. The line between the scrapbook of the home and the published periodical was further blurred by the use of the word *scrapbook* in the titles of several magazines: In the mid-nineteenth century, *Fisher's Drawing Room Scrap-Book* and *The American Scrap Book;* in the early twentieth century, Frank Munsey's *The Scrap Book,* and *The Official Bulletin and Scrap Book of the League of American Wheelmen,* among others. Other publications, like *Littell's Living Age,* announced themselves as digests and consisted largely or entirely of articles recirculated from other publications.

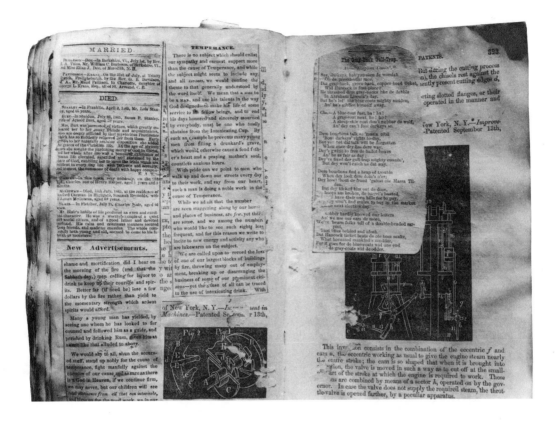

Scrapbooks might be shared with an intimate or large circle, and were a form of private publication, like the amateur press publications or "zines" of their day. They might be kept within a family as part of its library.

The occasional claim that the value of the scrapbook is transferable, and that it could be useful to others, is not universal. John Todd suggested that a published *Index Rerum* would be useful only to members of the same profession. But scrapbooks were often characterized as too personal and revealing to even be shown to others. The maker of the scrapbook carved an individual creation suited to individual interests out of mass-distributed material.

Scrapbooks and Classification

Articles on how to make scrapbooks, or describing the benefits of scrapbooks, repeat variations on the theme of the family—especially the farm family—sitting together in the evening making scrapbooks on their own favorite topics. In one such family, visited by Julia Colman, young Hubert, "is enamored of voyages and travels, and his big book has become almost a gazetteer. He arranges the articles alphabetically, and we often refer to them for information that we cannot find in books." The other children in this family have their scrapbooks on housekeeping, cookery, and poetry. There's Susie's "Guide to Health," and Charlie's "Horticultural Guide"; Hartley takes the mechanical and scientific items while "Libbie enjoys the pictures, and the baby glories in the wastepaper."[26] The classification of the clippings becomes inextricable from the classification of the children and creates value from the clippings. The scrapbooks become the common property of the family, consulted by all when in need of poetry or information about foreign lands.

Gurley praises a similar arrangement:

> Would it not be a pleasant picture of a farmer's family gathered around the fireside of a winter's evening overhauling a bundle of papers selected for this purpose? Each one may have his or her own book. The father may have the general topic of the "The Farm," the mother, The Home. One of the boys likes horses and cattle, and he is on the lookout for everything in that line; one of his sisters is poetical, another romantic, and "poetry" and "stories" will fall to them, and so on through all the topics.
>
> When the selections are made, let them be read aloud and discussed as to their merits and proper classification, and much good will will be gained by it. Where a custom like this is observed in a family, that home will be pleasanter, and that family will take an advanced standing in society.[27]

Scrapbook making is imagined as an occasion for domestic sociability, a way of uniting the interests of the farm and farmhouse through the medium of the pastepot. These accounts' emphasis on the farm home is a reminder that access to published material was not equally distributed. The ideal Gurley and Colman hold out is that by sifting through the detritus of the cheap press, the marginalized rural family can classify marginalized material and thereby create value from it. As they gather to pull apart mainstream culture and remake it for their own uses, they create their own cultural nexus, a knot of threads leading into and out of the family. They can make it even more meaningful and valuable by putting it in the form of a book, and even figuratively move it from the margin to the center of the page within the covers of an existing book. Old newspapers are clutter, but thematically organized volumes are useful.

But why did farm families have material that had to be remade in this way? Americans received quantities of cheap print because publishers—first newspaper publishers and then farm and family paper publishers—had discovered that subscribers could be exchanged for money via advertising and that high circulation, or at least the ability to claim it, brought advertising. Like the nineteenth-century patent medicine pitchmen who proposed wildly catholic and inclusive lists of ailments a single product purported to cure, hoping a more varied list of illnesses would appeal to more customers, many publishers aimed for high circulations by including something for everyone in the family. Such inclusiveness also made it more plausible for advertisers of varied wares to advertise in the publications. So cheap magazines and farm papers deliberately addressed multiple audiences, with columns on crops, animal husbandry, housekeeping, and religious matters; recipes, poems and stories; and amusements for children.

In practice, many scrapbooks are not so clearly systematized as Hubert's, Susie's, and their siblings', but contain a mix. Poetry, household hints, and columns of miscellaneous information in the same book are common. Makers of such scrapbooks appear to have chosen on the basis of taste, saving what they thought would be worth rereading. To the extent these books reflect categorization within the papers, they sharpen the categories by eliminating specific ones: collecting poetry, household hints, and varied anecdotes, but not crop information, for example. Scrapbook making gathered in and dispersed in a continual diastolic–systolic movement, and circulated labor and leisure as well: for adults, making them was both work and a leisure activity. Making them also occupied children who might otherwise require the adult labor of supervision. Scrapbooks were often suggested as an ideal gift for an invalid child or adult alone in the sickbed and in need of amusement.[28] The scrapbook absorbed material and labor, processing and transforming them.

Scrapbooks and Authorship

Although scrapbooks were understood as local productions, they sometimes had wider readership than the family. Some scrapbook makers, especially those who made classified sets with separate volumes on different topics, reported either lending them out, with particularly heavy calls for some volumes, or finding them so valuable that they allowed visitors to read them only on the premises. Scrapbooks were occasionally privately printed and circulated among friends, like a scrapbook "gathering songs and verses for over fifty years" that *Harper's Monthly*'s "Editor's Easy Chair" extols for its ability to summon the memories and tastes of an era,[29] or *An Old Scrap-book with Additions,* "printed but not published, for Distribution as a passing Token to Personal Friends" by John Murray Forbes, a railroad magnate.[30] Such anthologies moved out into the marketplace. The editor of a collection of poems and inspirational sayings explains, "For many years gems of poetry and prose have found their way into a scrap-book, until friends, who felt a delicacy in borrowing so well-worn a volume, have urged me to give it publication."[31] After Elbert Hubbard died in the sinking of the Lusitania in 1915, his Roycroft associates published his scrapbook, although his "storing up in his Scrap Book the fruits of other men's genius" had been only "gathering spiritual provisions for his own refreshment and delectation" and not intended for publication.[32] The genre continued into the twentieth century, including *The Dale Carnegie Scrapbook,* which intersperses Carnegie's own inspirational remarks with others he collected. And scrapbooks are frequently referred to as the source for anthologies intended for use by speakers and ministers, with such titles as *The Public Speaker's Scrapbook.* As in Bartlett, the original text may well drop out of view altogether, as an industry develops of collecting and recirculating pithy, pious, or inspiring sayings. Ministers and other speakers pick lines from such compendia and pass them around and collect them themselves, creating and supporting a genre of "quotes."

I have discussed ways in which web bookmarks or favorite web pages echo earlier scrapbook forms and uses, in creating coherence from broadly miscellaneous materials and materials without attribution, even if the coherence or associations are clear only to the creator. The web has also given new life and form to published commonplace books and scrapbooks. Dozens of sites collect and solicit "favorite quotes" or display the site owner's own collection. A few are specialized and serve as a focus for community ("A Commonplace Book for Advisors: A Treasury of Quotations for Academic Advisors and Other Mentors in Higher Education"); most are broad, uncategorized but searchable by keyword.[33] Like the scrapbooks and commonplace books self-published by wealthy men, web sites are usually created outside the constraints and citation conventions enforced by publishers and editors. On the web, source and author become further detached from the freefloating quotes. While authors' names might still be present, perhaps because they serve as warrant of the value of the passage, titles of works from which they came are more often omitted, or the source may be given as Bartlett's or another compilation. Written passages that were once part of a larger work enter into a half life as quotes appended as signature files to email as well, where their position suggests that they represent the uniqueness of the email writer's thoughts, rather than their original author's.[34]

As the fragmented selections recirculate further into new contexts, they may be pressed into incongruous service. George W. Bush exemplified this trend when he quoted a few sentences from George Eliot's *Adam Bede* in a 2001 speech at the National Summit on Fatherhood, to reinforce his praise of narrowly defined marriage as "incredibly important for children." He read:

What greater thing is there for two human souls than to feel that they are joined for life to strengthen each other in all labor, to rest on each other in all sorrow, to be one with each other in silent, unspeakable memories at the moment of last parting.[35]

While George Eliot's views on marriage were complex, her own twenty-four-year union with a married man who lived apart from his children and *Adam Bede*'s plot involving an abandoned pregnant woman who commits infanticide would suggest that someone who had encountered the lines in the context of the novel might not choose them to make this point. Bush or his speechwriters might have been lulled by the presence of the passage on over seventy-five websites for planning weddings, for documenting weddings, and for marriage counseling, several of which include the lines in wedding ceremonies. They may not have taken it from the "Marriage Builders" site, which misspells Eliot, adds the Hallmark phrase "to share with each other in all gladness," and omits the troubling "at the moment of last parting." (Marriage Builders lays down a firm line against unmarried cohabitation.) They probably did not take it from the personal "Spirit of Shaarilla" website which moves Marriage Builders' work a step further by adding the gladness line, lopping off the last parting, and attributing it to "Anonymous." The Osmond Network's "Valentine Community" makes the same changes and attributes the passage to Ranier Maria Rilke.[36] On the other hand, Bush's omission of a phrase in the original ("to minister to each other in all pain") hints at some other unlocated source.

The violence done to the passage sounds more like armed poaching than the more pacific-seeming gleaning, but de Certeau assumed his poaching readers actually read the work they'd mentally adapted to their own conceptions. This loose trade in detached quotations, gleaning from what other gleaners drop and then passing along the genetically modified products, may have accelerated with the web, but the practice has earlier roots. Nineteenth- and early twentieth-century scrapbook makers were invited to recirculate their work into others' anthologies as well. Papers even solicited, paid for, and reprinted the "priceless gems" readers had collected, in one case in a volume subtitled "The Old Scrapbook," which was thus "written" by 52,000 readers who sent in clippings as well by the authors of inspirational verse and anecdotes they clipped, initially for use in the *National Magazine,* and then in this anthology. The scrapbook was again linked to the heart as an alternative literary authority in this formula: "Those things that appeal to you must appeal to others; that note of inspiration laid aside—bring it forth and let us make a magazine that will speak the language of the heart as well as of the mind."[37] The second volume, subtitled "The Old Homestead," still makes the scrapbook an explicit source. Its contents come from "dear old grandfather or grandmother in serene old age, who with tremulous hands cut from their treasured scrap-books the selection that is to them a real 'heart throb' fraught with tender memories."[38]

One could participate in authorship without writing a line, in other words. Writing is understood as a process of recirculation, in which information is sorted and stockpiled until it can acquire value by being inserted into a new context. As often appears to be true on web pages, the origin of the material is less important than the new form it takes, re-sorted and made available in new ways.

A model in which the person who gathers the information is considered the author has late twentieth-century analogs, for example the daily newspaper reporter who calls in a disconnected story to someone who turns it into "writing" ("Gimme rewrite, sweet-heart"), while the reporter gets the byline. But it is striking that fiction writing, too, is sometimes the product of a good scrapbook. In a 1915 article in the *Writer,* "a monthly magazine of interest to all literary workers," Delsher Welch gives scrapbook keeping a ringing endorsement by invoking the example of Louisa May Alcott. Welch recalls her telling him, "The habit of reading with a pair of scissors in my hand has stood me in good stead for much of my literary work."[39] Alcott herself published a series of books with the overall title *Aunt Jo's Scrap-Bag,* metaphorizing stories as reused cloth, also material to be dealt with with scissors in hand. If even Louisa May Alcott was simply in the business of recirculating scrapbook clippings, or at least writers and potential scrapbook makers were invited to see her that way, then making scrapbooks could seem not so different from publishing a book. Locating and classifying the scraps made them as useful as other books. Putting them into the form of a book further asserted this authority.

Scrapbooks have a way of reifying the metaphors we use to think about reading. Per-haps it is the solidity of paper and glue that does it, more substantial than the metaphoric cut-and-paste functions of word processing programs. *Bricolage* and *construction* of mean-ing have become bywords for understanding how readers interact with texts, so much so that the concrete referent of these metaphors is nearly lost. But for nineteenth-century compilers of scrapbooks made up of clippings from newspapers and magazines, those terms are not metaphoric but straightforwardly descriptive. The scrapbook maker con-structs meaning by acting as a good bricoleur: collecting odds and ends and pasting down these assorted preexisting artifacts to create a new work, gleaning the matter of her own sustenance, improvising a new structure with them.

Scrapbook makers literally made new media out of old. These new media in some cases preserved and displayed the works the compiler most valued; in others they sorted overwhelming qualities of printed matter into usable selections, while they marked their makers' sense of affinity with the public world of books and their desire to participate in it. Making and keeping scrapbooks allowed participation in some of the functions of the author and publisher without lifting a pen, as compilers adapted writers' words to fit new and personal meanings. Like present-day web site builders, they adopted a fetishized selection of trappings of publication that appeared to grant authority as well. Through their new medium, scrapbook makers straddled the personal and the mass-circulated.

Notes

1. Henry T. Williams and "Daisy Eyebright," "The Value of a Scrap-Book," *Household Hints and Recipes* (Boston: Peoples Publishing Co., 1884), 132–133.

2. Margaret Lynn, *A Stepdaughter of the Prairie* (New York: Macmillan, 1914), 216–218.

3. Michel de Certeau, *The Practice of Everyday Life* (Berkeley: University of California Press, 1984), 174.

4. Henry Jenkins, *Textual Poachers: TV Fans and Participatory Culture* (New York: Routledge, 1992), 17.

5. "When you reap the harvest of your land, you are not to finish to the corners of your fields in harvesting, the full-gathering of your harvest you are not to gather; your vineyard you are not to glean, the break-off of your vineyard you are not to gather—rather, for the afflicted and for the sojourner you are to leave them," Leviticus 19:9–10.

6. Julia Colman, "Among the Scrap-books," *Ladies' Repository,* August 1873, 89.

7. Kate Sanborn, *Memories and Anecdotes* (New York: Putnam's, 1915), 4.

8. John Todd, Introduction to *Index Rerum; or, Index of Subjects; Intended as a Manual to Aid the Student and the Professional Man in Preparing Himself for Usefulness, with an Introduction, Illustrating its Utility and Method of Use,* 2d ed. (Northampton, Mass.: Bridgman and Childs, 1833, 1835), 6.

9. "Editor's Study," *Harper's New Monthly Magazine,* September 1892, 639.

10. Rufus Anderson, *Foreign Missions: Their Relations and Claims* (Boston: Congregational Publishing Society, 1874), 316.

11. John H. Holmes, an owner of the *Boston Herald,* quoted in "The Columbian Reading Union," *Catholic World,* September 1897, 864.

12. Anderson, *Foreign Missions,* 316.

13. Fanny Fern [Sara Payson Willis Parton], *Ruth Hall* (1854; New Brunswick: Rutgers University Press, 1986), 130, 158–159.

14. Lillie Devereux Blake, *Fettered for Life or Lord and Master: A Story of To-day* (1874, New York: Feminist Press, 1996), 259.

15. Jeannette Gilder, *Taken by Siege* (1886, New York: Scribners, 1897), 7.

16. See chapter 1, Ellen Gruber Garvey, *The Adman in the Parlor: Magazines and the Gendering of Consumer Culture, 1880s to 1910s* (New York: Oxford, 1996) for a discussion of gender and children's advertising trade card scrapbooks; other work is forthcoming.

17. "Melaia and Other Poems" (review), *Southern Literary Messenger,* March 1844, 8.

18. Alice Cary, *The Poetical Works of Alice and Phoebe Cary with a Memorial of Their Lives by Mary Clemmer* (New York: Hurd and Houghton), 45.

19. E. W. Gurley, *Scrap-books and How to Make them: Containing Full Instructions for Making a Complete and Systematic Set of Useful Books* (New York: Authors' Publishing Company, 1880), 15.

20. Lynn, *Stepdaughter,* 218–220

21. Ibid., 223, 230.

22. Ibid., 213.

23. Colman, "Among the Scrap-books," 90.

24. [Rounsevelle Wildman] "As Talked in the Sanctum by the Editor," *Overland Monthly and Out West Magazine,* July 1896, 4–6.

25. Scrapbook inside Richard Baxter, *Saint's Rest,* circa 1890, collection of the author.

26. Colman, "Among the Scrap-books."

27. Gurley, *Scrap-books,* 13.

28. See for example, "Sideboard for Children," *Ladies' Repository,* January 1864, 58.

29. "Editor's Easy Chair" *Harper's New Monthly Magazine* 69, no. 412 (September 1884): 630.

30. J. M. Forbes, *An Old Scrap-book with Additions* [Cambridge, Mass. University Press, 1884] "Printed but not published, for Distribution as a passing Token to Personal Friends."

31. *Between the Lights: Thoughts for the Quiet Hours.* The copy I saw was missing the copyright page and title page, but this is possibly the book of this title by Fanny Beulah Bates, published first by New York: A.D.F. Randolph, 1887, and republished in 1898 and 1899, the latter time by New York: T. Y. Crowell [c. 1899]. Unpaginated.

32. Foreword to *Elbert Hubbard's Scrapbook: Containing the Inspired and Inspiring Selections, Gathered during a Lifetime of Discriminating Reading for His Own Use* (New York: Wm H. Wise & Co.; Roycroft Distributers, 1923), Unpaginated.

33. Michael J. Leonard, "A Commonplace Book for Advisors: A Treasury of Quotations for Academic Advisors and Other Mentors in Higher Education," 1996–2001, Pennsylvania State University <http://www.psu.edu/dus/leonard/book/commtitl.htm> 2001.

34. For a longer discussion of this phenomenon, see Rachel Toor, "Commonplaces: From Quote Books to 'Sig' Files," *Chronicle of Higher Education,* May 25, 2001, B13.

35. Mary Leonard, "No Model for Marriage: Bush Writer Choice Was an Adulteress," *Boston Globe,* 9 June 2001, 4.

36. See, among many others, Wedserv. <http://www.wedserv.com/article2.asp?subID=226& articleID=2323> 2001. Other variants appear at David Schaff and Amy Lindahl, "The Vows" <http://www.ideazoo.com/wedding/vows.htm>; and at <http://members.aol.com/ bandrdesigns/wedding3.html>, where it becomes, "What greater thing is there for two human souls/ than to feel that they are joined together as one:/ To strengthen each other in all labor, / To console each other in all sorrow, /To share with each other in all gladness, / To be one with each other as they reflect back on the course of their life together." Marriage Builders. "Topic: In Honor of Sweetest Day—These Are Beautiful," posted October 21, 2000. <http://www. marriagebuilders.com/forum/Forum8/HTML/002234.html> 2001. It appears in the same form and with the same misspelling on "Victoriana's Romantic Love Collection," <http:// www.geocities.com/BourbonStreet/Bayou/6029/Rlove.html>.

37. Joe Mitchell Chapple, Foreword to *Heart Throbs in Prose and Verse: The Old Scrap Book* (New York: Grosset and Dunlap, 1905); front matter is unpaginated.

38. Joe Mitchell Chapple, Foreword to *More Heart Throbs: The Old Homestead.* (New York: Grosset and Dunlap, 1911), p. ii.

39. Deshler Welch, "Profitable Scrap-Book Making," *Writer* 27, no. 8 (August 1915): 113–115.

10 Media on Display: A Telegraphic History of Early American Cinema

Paul Young

For Bill Brown

It would be difficult to imagine the earliest complex story films made in the United States without the telephone and the telegraph. I mean "difficult to imagine" in two senses: Literally speaking, these technologies play prominent roles in many early films, including D.W. Griffith's short film *The Lonedale Operator* (American Mutoscope and Biograph, 1911), about an imperiled female telegraph operator; and in terms of the history of narrative editing, the plots of films like *The Lonedale Operator* would have been difficult for contemporary audiences to disentangle had the telegraph not provided justification for Griffith's crosscutting between one place and another. Tom Gunning compares the narrative role of *The Lonedale Operator's* telegraph to that of the telephones in Edwin S. Porter's lost film *Heard over the Phone* (Edison, 1908) and Griffith's *The Lonely Villa* (AMB, 1909) in that both the telegraphed message and the telephone call "tutor" the spectator to understand crosscutting as "a simultaneous hook up between distant spaces." In *The Lonedale Operator,* the intrepid operator (Blanche Sweet) uses her father's telegraph to contact a neighboring railroad station with the news that two drifters have descended on her station to abscond with a valuable mailbag (and likely with her, as well). Her frenetic tapping motivates the transition to a new shot and a heretofore unseen space—a neighboring station—for an audience used to single-shot scenes. Telegraphy thus became a metaphor for the power of cinematic crosscutting to expand the narrative's geographical scope and hasten its happy resolution. As Gunning shows, the film's editing demonstrates the powers over time and space wielded by both telegraphy and film, and it mobilizes the former as a technological model for imagining the capabilities of the latter.[1]

Telegraph and telephone also functioned to bring spectators into the narrative in new ways during these pivotal years. In a key article on the development of narrative editing, Raymond Bellour uses *The Lonedale Operator* to demonstrate that crosscutting was becoming synonymous with the very idea of cinematic narration by 1911. At the time, production companies were concerned to focus the spectator's attention squarely on the story and away from what were newly defined as distractions: shocks and spectacles of a violent, pornographic, or political nature that might disturb middle-class viewers and draw vocal reactions from audiences in poor and "ethnic" neighborhoods. With its unique power to delay the climax (will the engineer arrive in time to save the operator?) and ensure the audience's investment in the story, crosscutting was an attractive tool for reducing the amplitude of such shocks. For Bellour, Griffith's film predicts (if it does not fully articulate) the dominant mode of Hollywood cinema we now label *classicality,* which uses story-centered editing, varying camera distances, and other devices to focus our attention as seamlessly as possible on a unified ending and the values implied therein. In *The Lonedale Operator,* the reunion of operator and engineer boyfriend in a single shot after several minutes of crosscut suspense implicitly links narrative resolution to heterosexual union.[2]

But other values and other unions may also be at stake here, lodged in the literal and symbolic linkage of film to telegraphy, but visible only when those links are contextualized historically. Gunning demonstrates that at the thematic level, the use of telephones and telegraphs to generate suspense prompted audience identification by hinging that suspense on familiar utopian scenarios, and equally familiar anxieties, about electrical communication. If *The Lonedale Operator's* telegraph saves the day, the telephone in *Heard over the Phone* puts the protagonist in aural contact with his endangered loved one, then taunts him with his inability to reach her in time to save her from an intruder. *The Lonely Villa* remedies this distressing situation by concluding with its protagonist saving his wife and family in the nick of time, but not without first focusing on the frustrating immateriality of telephonic communication when thieves cut the very line that had enabled the husband to discover his family's endangerment.[3] The telephones of early narrative cinema often acted as screens for shared doubts about the social benefits of new technologies that banished old barriers of space and time, particularly when the protection of private spaces from public threats was at stake. By alternating views between public and private spaces like an all-seeing eye, the cinema figuratively borrowed the telephone's powers, rendering them visible and even spectacular; but it also foregrounded the vulnerability of both the private sphere and the technology that was supposed to provide domestic space with the ultimate lifeline.

If viewed primarily from the perspective of film's formal history or the turn-of-the-century discourse on the "terrors of technology," the telegraphs of early cinema seem practically interchangeable with its telephones. After all, the protagonist of *The Lonedale Operator* blocks the door of her office just as the wife and children block the parlor door in *The Lonely Villa,* coding the railway station as a space of similarly imperiled domesticity, or more specifically, endangered female virtue. But while Gunning is clearly correct to compare the technical and discursive functions of the two media in early films, telegraphy was always a much more public medium than the telephone. This important distinction suggests that even very similar cinematic exploitations of the two machines rehearsed quite different social problems associated with turn-of-the-century telecommunication. Our recently enriched knowledge of precinematic screen practice and the conventions of urban-industrial visual culture deserves to be matched by historically sensitive accounts of the electrical media discourses within which early films were produced and exhibited.[4] Although the telegraph had first been implemented in the United States in 1844 and was getting along in years by the time projected cinema entered the American scene in 1895–1897, films produced between the first years of cinema and the emergence of full-fledged classical film style around 1917 exploit telegraphy as if it were a novel medium.[5] The persistence of this "ancient" technology of modernity as a specific kind of mechanical icon—a machine that doesn't break down, one that preserves not only threatened individuals like the Lonedale operator but also bourgeois social order (the valuable mail pouch the operator protects is also saved)—leads me to postulate that such "demonstrations" of the telegraph helped early cinema to position itself as a certain kind of new medium, one that would resemble telegraphy in its public mode of address as well as in its powers over time and space. As such, the telegraph also provided a powerful fantasy image of the cinema's ability to link spectators into audiences at the local and national levels, via shared information and collective ideals of a nation joined by electricity.[6] Unlike the telephone, which projected an aura of impotence even at its most benign, the telegraph in early cinema was less a symbol for subjection to technological modernity than a figure for negotiating the ever-shifting relationship between public interest and private good.

Different as cinema and telegraphy were as technologies, turn-of-the-century discourse conceived of them as links in a chain of progress that drew the world more tightly together. Indeed, as both a thrilling new gadget and a carrier of messages—news, spectacles, stories, emotional and visceral effects—the cinema aspired to a place among "instantaneous" electrical media like the telegraph and telephone in the public imagination, and this positioning played a determinate role in the experience of early cinema. I will elaborate three areas in which early film invited comparison with the telegraph:

technological presentation and spectacle; news reportage; and filmic representations of the telegraph that addressed the changing definitions of time and space promoted by new media. Taken together, these aspects of cinema-telegraph relations provide a lightning sketch of a new visual medium that was resolutely self-conscious about its status as a communications medium. Tracing telegraphic discourse through the early American cinema, in fact, sketches out the foundations of a counter-history of the emergence of classical cinema—one in which the pleasures of classicality were founded not simply on the transparency of storytelling, as is sometimes argued, but equally and specifically on the pleasures of watching the cinema *work,* communicating information and arranging that information into a meaningful story.

Presentational Culture: Reality Made Strange

Tom Gunning has demonstrated that early cinema focused spectators' attention primarily on the technologies that produced moving pictures rather than the content of the images.[7] The Edison kinetograph and its rivals thus entered a presentational tradition of screened entertainments that was as old as the seventeenth century, yet thoroughly modern in its concern with mitigating the shocks of technological modernity, distilling those shocks into representations and claiming to present them scientifically.[8] Neil Harris refers to the nineteenth-century form of this tradition as the "operational aesthetic" of mass culture, in which new and unnerving sights were wrapped in the rhetoric of pedagogical demonstration.[9] New machines were great favorites among the objects thus presented, and demonstrators often displayed diverse machines in similar fashions, grouping different technologies together into a single show. The itinerant showman Lyman Howe began his career in 1883 by presenting his Miniature Automatic Coal Miner and Breaker to small crowds, then switched to the phonograph in 1890, and finally found his most successful niche when he added his homemade "animotiscope," a motion picture projector, to his phonograph concerts in 1896.[10]

Howe's shift from a miniature gear-and-girder show to displays of media machinery parallels Americans' increasing fascination with the array of electrical media available by the end of the century, each one more remarkable than the last, and all strengthening the promises of universal understanding and national unity that had been associated with the steam engine and the telegraph half a century earlier. Phonographic and telephonic "concerts" like Howe's were regular occurrences in the United States by the 1880s, providing the masses with an effective materialization of these promises. The concerts brought people together through technology, but did so the old-fashioned way—through shared curiosity, curiosity that led to social contact. And yet the media concert took place within the distinctly modern parameters of the distraction-seeking crowd and the spectacle of technological reproduction, a spectacle that could itself be replicated before other crowds elsewhere. These presentations offered an experience of democratization through technology that was explicitly hegemonic and consumerist, hailing the viewer-listener as a benefactor of new media's utopian potential and thereby cultivating more phonograph purchasers and telephone subscribers from among the amazed crowds. At the same time, however, these events were somewhat threatening to the elite class of "electrical experts" who provided the utopian electrical rhetoric upon which the concerts depended. Carolyn Marvin argues that public presentations of communications machinery were much more common in the United States than the experts tended to acknowledge.[11] Their silence on the subject of mass exhibition may have been a product of their concern to preserve their expertise from vulgarization, a concern that, as Marvin demonstrates, continually shows through in their derogatory tone toward the electrically uninitiated. Media spectacles offered a working understanding of these technologies to anyone who came to look and listen, and thus infringed upon the experts' source of authority, their (mystified) electrical knowledge.

The telegraph held a privileged spot in this culture of democratically dispensed (even democratizing) electrical spectacle. The first American demonstrations of telegraphic equipment, which David E. Nye dates to 1838, "brought excited crowds to the first telegraph offices, which often provided seating for spectators."[12] Audiences saw the telegraph work with their own eyes, but were nevertheless astonished at the results. Following Leo Marx, Nye refers to this epistemological break as the technological sublime, a "collective[ly] experience[d]," industrial analog to the Romantic sublime that Kant posited to describe humanity's relationship to the dangerous and awe-inspiring in Nature: "Instantaneous communication was literally dislocating, violating the sense of the possible." What made this experience akin to the Kantian sublime was the wonder that human industry instilled in audiences, the realization that "man [*sic*] had directly 'subjugated' matter" with communications technologies.[13]

This realization contributed only part of the "dislocating" effect, however. The *collectivity* of the experience of humanity's new authority over time and space was an equally important theme of media displays, and particularly displays of the telegraph. In the last half of the nineteenth century, public telegraphy demonstrations abounded in world's fairs and expositions, and in modernity's paeans to the democratizing powers of industry, as well as in more quotidian venues. Born of the need to create markets and entice future technical laborers, as well as to reduce anxiety over the rapid changes brought by the Industrial Revolution, technological expositions undertook to "explain, educate, and interest the people in the new artifacts."[14] Technology historian K. G. Beauchamp reports that telegraphy was on grand display as early as the London Great Exhibition of 1851. Its appearance there was spurred by intense interest in the transatlantic cable then under development (although the cable would not become a reality until 1858). The telegraph's status as a popular attraction rested in its theoretical ability to expand the community outward from the microcosmic crowd of a world's fair into the world itself. But in the context of popular exhibitions, the telegraph was also exploitable in another democratic register, that of mass culture spectacle for its own sake, and the antiauthoritarian vulgarity that accompanied it. Some displays intermingled the rhetoric of the technological sublime with conventions of mass amusement, turning the telegraph's formidable power over space and time into a noncondescending entertainment of a sort familiar to the leisured masses. A remarkable example of this promise is the "comic telegraph" displayed at the Great Exhibition, which consisted of an effigy of a man's head framed by a wooden box; the mouth of the dummy "moved meaninglessly" as small flags above the head were exchanged with each other by electrical remote control. Such "trumpery and trash" drew complaints from electrical experts and other critics of the exhibition who were perplexed by the large number of exhibits intended only to "evoke wonder from the visiting population" rather than introduce them to machines with public or private utility. Of course, not every telegraphic exhibit had to work so hard to entertain. Various telegraphic and telephonic devices, from printing telegraphs to copying telegraphs (facsimile machines that transmitted handwritten signatures) to live telephone concerts and wireless telegraphy drew delighted crowds at expositions and world's fairs at least through the St. Louis exposition of 1904, without such mass cultural trappings.[15] But the "comic telegraph" suggests a tradition of antiauthoritarian humor in media display that operated around, and in tandem with, the operational aesthetic. This "low" tradition undercut the inflated importance of telegraphic discourse and challenged its projections of universal understanding and communications access. The anonymous babbling head of the dummy could have been understood by different audience members as a typical person, an electrical expert, a telegraph official, a politician, or some other authority figure with a level of access to telegraphy not common in either Britain or America. Whoever he was supposed to be, the dummy mouthed the sound and fury of the new machine but signified nothing, perhaps implying to less enthusiastic spectators that telegraphic discourses were hollow, the exciting messages sent over the "lightning lines" a showy front for a technology that merely meant business as usual for capitalism, class structure, and politics.

When projected cinema entered this world of technological amusements around 1895, it achieved a synergy between communication and amusement comparable to that suggested by the comic telegraph. For one thing, the cinema not only demonstrated the globe-encircling powers of new media; it *showed* them at work. Charles Musser and Carol Nelson establish the continuity between cinematic displays and presentations of other media in their discussion of Lyman Howe's switch from showing the phonograph to showing films: "As Howe had earlier done with the phonograph, the outside world was brought inside and audiences saw one distant place after another transported before their eyes, new significance given to the ordinary."[16] If the phonograph concert disembodied and recontextualized sounds made elsewhere and "elsewhen," telegraphic display had been showcasing such mediated "presence" in similarly uncanny fashion for some time. The cinema technically resembled phonography more than telegraphy in that its representations were prerecorded and apparently quite literal compared to the telegraph's encoded messages, but all three media offered audiences a distinct feeling of spatial and temporal dislocation.[17]

Telegraphy displays, I am suggesting, helped prepare the way for what Gunning calls the cinematic "aesthetic of astonishment" by making visual entertainment out of new technology's spatiotemporal powers. Gunning describes this aesthetic as a mode of presentation that never concealed illusionistic intent (in keeping with the operational aesthetic) and yet produced a sense of awe in the audience at the very fact of the trick.[18] No explanation could fully cushion spectators against the discomfort produced by films like the Lumières' infamous *L'Arrivée d'un train* (France, 1896), in which a train rushed toward the camera in three-quarter view to the consternation of its viewers, but this was a distinctly modern discomfort born of being positioned between a metaphysical sensibility regarding new technological phenomena (we might distill this sensibility into a sentence: "This feat being performed, this simulation, was simply not possible before") and a sophisticated, incredulous attitude toward the previously impossible that resulted from the operational aesthetic's pedagogical impulse. Beauchamp and Walter Benjamin both characterize industrial modernity in terms of the production of a spectatorial gaze that imparts to its subjects the sophistication that capitalism requires of them—sophistication as consumers, laborers, and members of the progressive community that will benefit from such technological marvels—while at the same time retaining the sense of wonder crucial to making technologies and other commodities desirable in early consumer capitalism; one must be allowed to attribute near-magical qualities to commodities if one is to participate in the phantasmagoric pleasures of modernity. Benjamin writes that "the framework of the entertainment industry has not yet been formed" by 1798, the year of the first Parisian industrial exhibition, but "the popular festival supplies it."[19] This framework, as Benjamin indicates throughout his work on mass culture and technology, depends increasingly on entertainment media that carry the dialectic of consumerist regimentation and distracted fantasy into the everyday world outside the exposition gates.

The disjunctive, uncanny quality of the technological display has one more facet that links telegraphy to early cinema: a flavor of the occult. Gunning argues elsewhere that the notion of duplication was at least as important to the early reception of still photography as was the idea that such images were "capable of presenting facts in all their positivity and uniqueness," and that duplication lay at the center of spiritualist and occult fascination with photographs, serving as the very source of their uncanniness: They seemed to undermine identity on the one hand by producing copies of the unique individual, and cheat even death and nature on the other by allowing a perfect image to outlive its subject.[20] Early films, especially the works of magician Georges Méliès, continued to locate an affinity between (cinematic) photography and the metaphysical, both in the profilmic events they recorded and in the cinematic tricks they played with, and on, reality. In a revealing article, Richard J. Noakes argues that the occult played a similarly important role in the initial discourse about telegraphy, specifically regarding the sublime impossibility of closing the gaps of space and time. In Victorian England, Noakes writes, "both telegraphic and spiritualistic forms of communication proved troublesome" to rational modern subjects, and "promoters of either scheme could be accused of fraud, ignorance, and over-credulity." British railroad interests initially refused developer John Frederick Daniell's offer to wire the railway lines because his promises smacked of the occult.[21] In effect, to most bystanders it seemed as unlikely—or likely—that one could "duplicate" oneself electrically by communicating with other living beings via the disembodied "lightning lines" as it was that one could communicate with the dead.[22]

In both cinematic and telegraphic display, then, an unsettling new sense of presence was at stake, which Jeffrey Sconce characterizes as "a fantastic splitting of mind and body in the cultural imagination," a concept that found its first modern expression in telegraphy.[23] The operational aesthetic debunked the metaphysical aura floating about each medium, only to make even the most sophisticated audiences rub their eyes in pleasurable disbelief at this sublime technology. Since telegraphy and photography both emerged in the 1830s, it may be fair to suggest that the two jointly enabled an occult sensibility regarding media communication. When the most realistic of visual media was put into motion by the cinema, the historical relationship between photographic and telegraphic occultism was revivified: Moving pictures made a photographic spectacle out of the distinctly modern instantaneity and disturbing sense of presence that kept the telegraph a device worthy of public fascination even at century's close.

Reporting the National News

The telegraph was expected from the beginning to excite its users and audiences with its aura of dislocated presence, but its most heavily promoted social promise was to unite those audiences via information. American lawmakers and telegraph executives, the main proponents of this promise, echoed democratic rhetoric about the railroad that dated back to Daniel Webster's occasional speeches of the 1830s and 1840s.[24] Webster's speeches posited a railway that automatically stitched the nation into a whole, loosening the bonds of American regionalism through steam-powered trade and travel. In similar fashion, champions of telegraphy extolled its power to "achieve and maintain national coherence."[25] In a prognostication typical for the time, George Prescott, the superintendent of electric telegraph lines for Western Union, wrote in the 1866 edition of his 1860 treatise on the telegraph that "its network shall spread through every village, bringing all parts of our republic into the closest and most intimate relations of friendship and interest."[26] In practice, however, access to the medium was hard to come by. Despite its initial use as "an instantaneous two-way medium" and predictions of private, individual utility made by no less an authority than Samuel Morse, the American telegraph was quickly developed and regulated (by champions of capitalism like Prescott himself) as a machine for doing business and little else. Everyone but the richest of electrical experts and businessmen paid often exorbitant rates to telegraph monopolies like Western Union for the privilege of sending brief, semiprivate messages.[27]

One powerful way in which the telegraph *did* seem able to fulfill its democratic promise, even under these economic conditions, was in its ability to transmit the news to an eager public. Telegraphic news missives usually reached the public in the form of newspaper stories transmitted by Western Union and/or packaged by the New York Associated Press (later the AP), which was formed in 1848 to distribute telegraphed news stories to a cartel of member newspapers.[28] By informing people of events that took place hundreds or thousands of miles away, telegraphic news services lent an electrical spark of excitement to the process of nationalizing American culture. Readers who digested these stories as quickly as the newspapers printed them could feel personally engaged with events of national importance almost as quickly as they occurred.

But telegraphed news did not confine itself to the papers; it also went on public display to deliver stories as they unfolded, preparing the way for cinema's own renditions of the news. Telegraphy-fueled "performances" of news dramatically staged events of public interest in real time, introducing a pointedly narrative element to the aesthetic of telegraphic display. During the Civil War, war reports were regularly transmitted via telegraph and posted for public consumption, a practice that proved the existence of a nation-wide demand for the speedy and universal delivery of news. On the eve of the 1896 presidential election, just as urban vaudeville houses and other theaters in the United States began to exhibit moving pictures, telegraphed and telephoned poll returns were displayed to waiting crowds via written cards, search lights, and stereopticon slides. Other spectacular "telegraphic" displays delivered content more in keeping with their mass-cultural mode of address. Between 1884 and 1894, at least three American concerns used telegraphy to receive play-by-play information about baseball games, which was then presented to assembled fans by way of creative devices like model baseball diamonds covered with nametags and playing pieces that could be moved to represent plays.[29] In each of these cases, the emphasis of the spectacle rested not on the attraction of the technology alone, but on the unfolding of a series of real events occurring elsewhere; in other words, the point of the display is the representation of a suspenseful narrative, the baseball game. Mark Twain's short story "Mrs. McWilliams and the Lightning" (1880) humorously ties together the political, spectacular, and textual threads of such telegraphic reportage by recounting a long and worrying lightning storm that turned out to be no lightning storm at all, but cannon fire in response to an important telegraphic message: "Garfield's nominated—and that's what's the matter!" Twain manages to criticize modern rationality, telegraphy, amusement-style "reporting," and future president Garfield himself in the space of six pages, but the "lightning" of the title singles out the telegraph for special scrutiny ("lightning lines" being a popular nickname for the telegraph). Perhaps Twain is equating all of the above developments with being struck by lightning, and holding the telegraph responsible for facilitating long-distance disturbances of the private and regional peace.[30]

Like the electrical media shows that Twain satirizes, filmed realities or actualities (usually single-shot films of people performing everyday activities or quasi-documentaries of famous or exotic places) offered both an impression of presence and a certain textual pleasure to their spectators. In their search for commercial material, filmmakers like Edwin S. Porter began to stage current events for their films by 1901 and helped create a demand for timely films of a more quotidian variety than the spectacular (and often staged) actualities made during the Spanish-American War and "Cuban crisis" a few years earlier. Porter and George S. Fleming's *Kansas Saloon Smashers* (1901), for example, turned the news into both spectacle and text by reenacting—and sometimes parodying—recent news stories (in this case, Carrie Nation's destruction of saloons in the name of temperance).[31] The concept of cinema as a "visual newspaper," as Robert C. Allen calls it, tied the new medium to newspaper conventions of visualization such as front-page engravings or political cartoons, and also to the rapid delivery of stories everyone was talking about, imbuing the cinema with an immediacy similar to that which telegraphy had made possible in the press.[32]

In at least one suggestive case, telegraphic news and cinema overtly overlapped during performances. In 1896, a Broadway theater interspersed slides reporting national election returns with a series of Vitascope films. The crowd that had gathered to watch the returns cheered the latter loudly,[33] in spite of the fact that the views may have had nothing to do with the election (though Biograph films of McKinley did circulate during the 1896 campaign). Placing the "reports" produced by these two media together at a tense moment like election night probably reinforced the sense of disembodied presence associated with each medium. In particular, the brand-new medium that sparked photographs to life would seem to have gained a further infusion of instantaneity and urgency from being placed next to telegraphed reports. Jonathan Auerbach shows that audiences who saw Biograph's *McKinley at Home—Canton Ohio* (Bitzer, 1896) at different stages of the campaign read different meanings into the candidate's on-screen actions, in particular his reading of a telegram. No matter when the film was shown, it seemed to portray the candidate's response to a specific current event, whether an upturn in McKinley's support or a report of his victory. Auerbach calls this phenomenon "news with no content—or whose content varies according to the moment of its screening." Auerbach's essay suggests how closely related the notion of films as "realities" was to a concept of cinema as *live,* an implication further supported by occasional convergences with the telegraph: "By deliberately incorporating into its drama a prior medium of mass communication, the telegram, . . . [*McKinley at Home*] self-consciously signals its own power to deliver electrifying messages across time and space."[34]

Following Gunning's argument about the incredulity of early audiences, I do not claim that viewers believed that what they saw on the screen was capable of interacting with them or that they mistook the projected "events" for events that were taking place in real time. I do think it possible, however, that technologically sophisticated viewers, charged with excitement about electrical instantaneity and intrigued by cinema's uniquely visual position among new media, could gladly and self-consciously suspend their disbelief and react to certain views as if the events portrayed were happening at the moment of exhibition, via a pretelevision version of closed-circuit viewing (in fact, devices for "see[ing] by telegraph" were already predicted by 1889).[35] Newsworthy events like a Chinese man being taken into police custody in San Francisco, the aftermath of the 1900 Galveston flood, and the Russo-Japanese peace conference of 1905 traveled all over the country by film, becoming the closest thing to "seeing by telegraph" available to early film spectators, who found this function entertaining if not indispensable. Musser even credits the Spanish-American War with saving moving pictures after their novelty had begun to wear off.[36] If only in their imaginations, early cinema's audiences found a constant stream of opportunities to take advantage of the promise of national unity offered by both the familiar wire services and the new medium of film.

But did the news delivered by this utopian medium succeed in unifying its viewers? And to what degree did visual "transmissions" of the news live up to the democratic promises of communications technologies? Before attempting an answer, we should recognize that these are two distinct questions that turn-of-the-century progressive rhetoric tended to conflate by equating shared knowledge with national solidarity. Telegraphy had symbolized the inevitability of nationwide communication and understanding for decades before the advent of film, but the claim that the telegraph would "unify" the nation rested solely in its delivery of the same news stories throughout the country. The telegraph itself opened no literal opportunity for individuals to respond to or debate the stories reported. The privatization of the American telegraph network led quickly to Western Union's so-called natural monopoly, a situation communications insiders characterized as "democratic" in contrast to Europe's state-owned, state–regulated networks. As George Prescott put it in 1866, "[Telegraphy] is alike open to all, and telegraphic despatches [sic] are 'household words' among the poorer as well as the wealthier citizens." In fact, however, private control helped guarantee a "monopolistic control of knowledge" along with economic control over the network: "Whereas the mailbag could carry numerous communications simultaneously, the telegraph transmitted only a single copy of one message at a time. The telegraph line thus promoted equality of knowledge across space, at the price of a monopoly for certain messages in time."[37] These "certain messages" were the news stories that Western Union and the AP delivered to hundreds of newspapers nationwide. No matter how biased the stories were, they carried the prestige of telegraphic instantaneity and objectivity, and were generally the only versions of the stories available on a nationwide basis. National unity, the end to interregional squabbling that the train, the telegraph, the telephone, and the cinema were all expected to bring, was a myth of industrial-age transport and communication that disguised the enforcement of political consent as a means to full participation in democracy.

The potential for cinema to exploit its own "objective" character for political purposes was just as great. Even the amusing and overtly biased films of "lightning artists" sketching political figures to their own specifications[38] carried the palpable authority of photographic realism. To make matters more complex, filmed information flowed more decisively in one direction than even that of telegraphy, offering not even the theoretical possibility of responding to the filmmakers through the medium. The concomitance of these factors might lead us to conclude that the masses were serially fooled by war or election films into buying politicians, and politics, that they didn't want and may have ill-served them. Auerbach appears to espouse this view when he compares the effects of what he calls "highly calculated" early films of news events to the insidious effects of news monopolies on news reporting, arguing that news as reported by films like *McKinley at Home* would "have left some but not a lot of room for its viewers to reclaim meaning on their own terms."[39]

The fact that news reportage in early cinema discursively allied itself with the authoritarian "positivism" of the telegraph, however, does not mean that the films *succeeded* at producing the consent they may have been intended to produce. The historical ambiguities surrounding *McKinley at Home,* its different meanings for spectators at different times and even the multiple titles used by both Biograph and exhibitors when promoting the film, all suggest how difficult it was for early filmmakers to assign unambiguous significance to such films. For one thing, any film constructed to induce certain responses would have had to contend with the same unpredictable exhibition practices met by fiction and trick films. Exhibitors had the freedom to place the McKinley film in an unflattering position on the program, or to deflate its effect through ironic commentary or music.[40] Where Auerbach sees absorbed spectators identifying with their political champion, I see audiences for whom the discursive centrality of this film was far from clear, appearing as it inevitably did among many other short films that were rarely political in theme, judging from descriptions given by contemporary press accounts (reprinted in Kemp R. Niver's collection *Biograph Bulletins*).[41] One account describes even McKinley's supporters as "fun-loving" in their waves to the president and playful demands for a speech. Other reviews make little distinction between McKinley's image and all the other cinematic views on the program, some breathtaking (the Empire State Express rushing toward the camera), others amusing ("the drinking scene from 'Rip Van Winkle'")— they were all, in a word, "astonishing."[42]

Even the most acutely partisan responses to McKinley were ultimately out of the producer's control, as Auerbach's own examples demonstrate. He describes a woman who "insisted upon making a speech" when watching the film as an empty vessel channeling the "mute" candidate on the screen into the theater,[43] but whatever her relationship to McKinley's message, she was also taking advantage of the nonhierarchical relationship between screen, spectator, and audience that characterized cinematic exhibition at the time. Early film demonstrations offered viewers something that the telegraph by 1896 only granted in nostalgic fantasy: a public platform for response, not to the film producers or to the people appearing in the pictures, but to other patrons who viewed the "news" events (real or reconstructed) on the screen. Given the distracting, carnivalesque circumstances of cinema's first years, it is extremely unlikely that viewers would, or could, have become so absorbed in *McKinley at Home* that they fantasized themselves "at home" with McKinley, as Auerbach claims. If anything, the film's presentation of McKinley at home makes a point of publicizing the domestic sphere, both for McKinley and for the audiences, rather than domesticating a public experience of the candidate. Vaudeville theaters and other early venues like New York's Eden Musée or Koster and Bial's were scarcely like "home" for their patrons. In fact, an important attraction of moving picture shows was that they brought their viewers *out* of the home, away from paternalistic and gendered restrictions on behavior, and into contact with the neighborhood and the possibilities that public amusements offered for liaisons, conversations, and distraction from labor and familial duties.[44]

What must be recalled, in the case of telegraphic and cinematic news alike, is the fact that neither technology nor its socioeconomic institutionalization can completely determine the meaning of the images or messages transmitted. Indeed, not only films but also telegraphed news stories were suspected of bias almost from the beginning. Despite central "processing" of the information it communicated, the telegraph, in Harold Innes's words, actually "accelerated the process of political fragmentation in the United States" because it was owned by private interests which used the network to reinforce local authority over such politically charged issues as banking and labor.[45] Reportage of the same news to all corners of the country—especially federal policy disputes—sparked varied feelings depending on the region and on which newspaper packaged the story for which readership. And, perhaps most important considering the discursive conflation of telegraphy with open political discourse, telegraphic news permanently transformed the American political scene by opening up federal debates, formerly kept private by elected officials in Washington, to "national public opinion." The telegraph thus rendered impossible the politician's trick old of speaking in a "different voice" to different regions, an underhanded but common maneuver that had actually helped to preserve the Union before the Civil War.[46] Telegraphic news transmission troubled unselfconscious national unity by making regional positions regarding slavery, states' rights, and other divisive issues impossible to ignore. Central control exerted over news reportage, then, was no match for regional and cultural differences.

As news "networks" providing what in practice became "news with no content," to return to Auerbach's helpful phrase, telegraphy and cinema jointly expanded the interpretive powers of their audiences into political territory. Reviewing visual newspaper films within the contexts of mass-cultural presentation and telegraphic news reporting reduces the temptation to overstate the power of the "objective" text. Newspapers and films alike often acknowledged their own editorial functions, presenting them as correctives to the equally biased practices that influenced AP reports, while those films that were not explicitly biased were subject to interpretation by exhibitors. These interpretations were in turn scrutinized by viewers who quickly learned, under the solicitations of early consumer capitalism and the emergence of a nationalized political scene, that their opinions counted for something. Previously distinctive groups of voters and readers were becoming the mass-mediated, mass-consuming public to whom manufacturers, and increasingly politicians, had to appeal if they expected to stay in business. Instead of narrowing and unifying responses to the news, the telegraph helped provide the conditions for the emergence of an audience that could identify itself as "national" in what it consumed, but still reserved the right to consume the texts of mass culture on its own, local terms.

Presentation / Narration: Projections of the Telegraph in Early Film

I return now to representations of telegraphy in early cinema, but this time with a sense of the complementary discourses of these technologies, and particularly the routes to national community offered by each—the electrical-political dimension, in other words, of the geographical leaps motivated by the appearance of the telegraph in *The Lonedale Operator*. We can learn much about the attraction of telegraphy by comparing telegraph films to telephone films, but a distinction needs to be made between the cultural meanings of the telephone and the telegraph in order to enter the films' spectatorial context. The telephone was a newer medium with even greater utopian potential, and greater astonishment value, than the telegraph. First successfully demonstrated in 1876, often in staged "concerts," it communicated individual voices directly to the listener, bypassing (it seemed) the technical and economic "necessities" of leaving transmission and reception up to trained encoders and receivers. In practice, however, the telephone divided the economic classes in ways that were even more obvious in quotidian life than the strict line between businesses and citizenry drawn by the telegraph industry. Private users could get direct access to phones, but no matter how ubiquitous public telephones were at the turn of the century, having a private phone in one's home was an economic privilege, not a right. Whereas telegraphs were at least housed in public offices, telephones were privately leased by the economic elite, and rapidly became a mark of middle-class distinction. Indeed, the Bell company reinforced this distinction by attempting to regulate the use of phones by nonsubscribers through various penalty schemes—a strategy actively resisted by "emerging networks of telephone sociability" in which subscribers regularly allowed neighbors to borrow the phone.[47]

An early American gag film, *The Telephone* (Edison, 1898), rehearses working-class ambivalence about this pricey private medium. A man stands in front of a hand-crank telephone marked with a sign: "Don't Travel! Use Telephone! You can get anything you want." He then takes what appears to be a glass of beer out of the telephone cabinet, and the film ends. This crude but effective gag characterizes early cinema's fascination with electrical media quite well, for it exploits the device as a technological spectacle, and at the same time uses it to spark a comic strip-style narrative, here a knowing joke at the expense of American technological utopianism. Fantasies ran rampant at the time that the telephone would eliminate the need for travel, allowing the bourgeois citizen to command the bounty of mass-production capitalism from the comfort of home, or at least a private booth (it's impossible to tell which is represented by the blank sets of this film, though we might assume from the presence of the advertising sign that this is a saloon, a public place).

The Telephone, however, acknowledges the contradictions in this promise in multiple registers, in keeping with the "ambivalence" of address and meaning that set early cinema apart from classical narrative cinema.[48] The magical materialization of the beer mocks the overstatements of electrical media optimism, for the working-class audiences of early films would have been as likely to see a real telephone deliver them beer as to see it deliver on its promises of national unity or the eradication of class difference. Even if part of the population could cordon itself off and use only the telephone to venture into public (as in the media fantasies cataloged by Marvin), that part would still depend on a service class to cater to its needs. I suggest that the humor of this gag fundamentally depends on the bitter recognition that in 1898, this telephone fantasy is just that, a fantasy: an unrealizable dream that the telephone could offer anyone of any class an elite position protected from even awareness of labor. But the film retains a utopian glint by visualizing the medium's bounty as a *beer*. The mug of beer aligns the film firmly with a male working class that cherished beer as an icon of that vital locus of urban social exchange, the saloon.[49] The private medium showcased by *The Telephone* delivers not merely a beer, but an avatar of publicness at its most disorderly and unsanctioned. The medium delivering this ironic visual message about the telephone thereby sends a somewhat different message about itself: Film does not offer its patrons false technological promises or upper-class conveniences, but jokes at the expense of both—a sign of recognition of, and solidarity with, viewers unlikely to have phones installed in their parlors.

The telegraph in early cinema, on the other hand, seems more clearly cast as a "people's" medium, not because individuals controlled it in the world outside the theater (the opposite was the case, as I have discussed, although telephones had only dented the private market by 1895), but because its real social functions were more obviously public in nature. The telegraph had been developed into a technology for "high-speed, one-way communications" sent in the order received (with business and news agencies given top priority), rather than a private or even a point-to-point medium.[50] Perhaps for this reason, the stories generated about it in early films tended to involve social authorities with access to the technology (railroad operators, firemen and policemen), and events the consequences of which reached beyond the concerns of individual households (disasters, crimes, and other events with public impact). In spite of its status as a monopolized technology, the telegraph in early cinema plays the part of the public medium's public medium—a device similar to cinema in its everyday relationship to the masses as masses, and thus a technology onto which the cinema projected images of its own evolving relationship to its spectators.

Two of Porter's pioneering story films involve telegraphs playing public roles of this sort: *The Great Train Robbery* (1903) and *The Life of an American Fireman* (1902–1903). *The Great Train Robbery* centers on the telegraph first to communicate the depths of the criminal gang's audacity when they bind and gag a railroad telegraph operator, then to help motivate the climax of the film by showing the operator "telegraph[ing] for assistance by manipulating the key with his chin," as recounted in the 1904 Edison catalog.[51] Intriguingly, the telegraphed missive apparently has little effect, and the operator has to rouse the posse in person. Since the long shot of the telegraph office does not privilege the device, the telegraphic chin business alone would have been insufficient motivation for the appearance of the cowboys; rather, its function seems primarily demonstrative and symbolic. The telegraph adds to the social urgency of the robbery by buttressing the film's theme of the cause-effect relationship between reportage of events and the restoration of social order, even if the telegraph does not literally perform such a function in the plot. The telegraph operator's incapacitation symbolizes not just personal crisis but social chaos; suitably, then, his return to duty is the film's first step toward the robbers' incarceration and the return to social order.

In *The Life of an American Fireman*, Porter shoulders the telegraph with much more central visual and narrational functions, and materializes its social promises more emphatically. The second shot of the film is an insert of a fire alarm telegraph, clearly marked as such, being pulled by a hand reaching into the shot from offscreen (figure 10.1). Here Porter showcases the telegraph for its sensationalistic value, continuing the tradition of telegraphic display in a new medium; the alarm device was still novel enough by 1903 to provoke curiosity and stimulate discussion.[52] We could easily mistake this shot for a precocious use of an electrical medium to implement something like classical narration, since the alarm motivates the expansion of story space into the fire station, where the firemen rouse themselves to answer the call. However, I suggest instead that the startling cuts from the fireman's dream vignette in the opening shot (figure 10.2), to the alarm, and finally to the fire station have perhaps less in common with similar series in *The Lonedale Operator* than with the serial presentations of baseball plays in telegraphed game displays. Baseball displays allowed audiences to assemble a simple narrative—the game—out of the series of plays received and presented by the operator, but because the medium's power over space and time took center stage, the causal chain within the game between play A and play B competed for attention with the causal chain between the telegraph's operation and the presentation of its message. In similar fashion, *The Life of an American Fireman* creates a causal link between event A (the alarm being pulled) and event B (the firemen responding to the alarm by getting ready) while foregrounding media transmission as an equally exciting narrative, but this time the kinetograph is the medium whose power demands the most attention. Displaying the telegraph prominently as a conceptual touchstone, Porter delivers images of the alarm technology doing its stuff, then moves into a spectacle of urban disaster and movement that only cinema could offer to an audience at a technology demonstration, a visual recreation of the live "Fighting the Flames" show that astonished thousands at the Paris Exposition of 1900 (and arrived at Coney Island the year after Porter's film was released).[53] *The Life of an American Fireman*

Figure 10.1   *Life of an American Fireman* still.

effectively positions the cinema in the telegraph's old place in electrical display culture, a technology that gathers viewers together to ogle its spatiotemporal prowess.

The discourse of the telegraph and kinetograph as communalizing forces intersects with another theme that *The Life of an American Fireman* shares with *The Great Train Robbery,* that of the connection between private sentiment and public action. The film begins with an image of a fireman in repose (see figure 10.2), possibly dreaming of his wife and child as they appear in a matte "bubble" next to his head (the film leaves the identity of the woman and child ambiguous, even deliberately so, as the Edison Company's speculative description of *Life* further attests).[54] Porter refuses to individualize him any further, however; the title casts him simply as an "American." After the close-up of the alarm, the fireman must move outward from his private reverie and into public duty, a move the new telegraphic alarm promoted and enforced in American cities, and equally a move that the *shot* of the alarm motivates in this film. The shot changes from fireman at rest to tele-

Figure 10.2   *Life of an American Fireman* still.

graphic alarm to firemen in action builds on the technological analogy between telegraph and cinema by placing the spectators in a relationship to the film that parallels the fireman's relationship to the alarm: Both machines not only transmit information about one space into another space, but also bring private subjects out into the open, turning them from singular figures with individual dreams to participants in an idealized "American" experience, broadly defined as everyday heroism for the fireman, and sensationalistic technological amusement for the audience. In its implications of the cinema's power to collectivize and nationalize the otherwise anonymous masses, *The Life of an American Fireman* borrows fantasies of democracy through telegraphy that were more than half a century old.

I must stress, however, that the film's discourse of nationalization is more a shape left to be filled out by subjects and practices than a doctrine determining the nature of a political subjectivity. *Life* begins with a reassuring image of middle-class values sitting at the helm of public safety as the fireman dreams of domestic tranquility, but even as it returns to this theme when the fireman rescues a mother and daughter (perhaps the same figures portrayed in the fireman's daydream), it does so amid terrific narrative ambiguity, and only after a race to the rescue which emphasizes the thrills of spectacle and speed for their own sake.[55] Here the cinema symbolically delivers on a promise that the Associated Press made to its readers but never quite kept: the promise of "raw" news, the unbiased reportage of events. The film's putative themes of public duty and technological discipline are presented so ambiguously that its second shot would likely have had a much less positive meaning for actual firemen than for the average thrill-seeking film viewer. To firemen across the country, and particularly in cities like New York and Philadelphia, the alarm telegraph signified the destruction of the municipal fire department system and the grassroots political force it represented. City firemen, who were "frequently the fomenters of urban riots . . . viewed technological innovation as threatening to their numbers and hence to their existence," and resisted full deployment of telegraphic alarms for decades before finally succumbing to the downsizing, efficiency, and professionalization that the alarm system represented; as late as 1902, a census report found that firemen were the "bitterest enemies" of the alarms.[56]

Even though all the events presented by Porter's film are fictions staged for the camera, the ambiguities of relationships among characters, the impersonality of the long shots showing the race of the horse-drawn fire trucks, and the infamous closing sequence which shows the fire rescue in its entirety from two different locations in two separate long takes, all left audiences ample room to interpret the story content and the temporality of the events shown.[57] Certain viewers might even have taken the telegraph to task for helping put the volunteer out of work in the name of social order and efficiency, and taken offense at the very existence of this film. As Musser points out, firefighting films took the power to represent firemen and their tasks out of the hands of the volunteers who willingly posed for films like *The Life of an American Fireman,* the same volunteers who were losing their prominent positions thanks to the increasingly bureaucratic and centralized mode of American capitalism and social organization that cinema simultaneously depended upon and symbolized.[58] Porter's film is not a news report, but it produces at the structural level a mandate that audiences arrive at their own interpretations, whether consenting or dissenting, of what transpires on the screen—the ideal position of the news consumer in a democratic society.

Conclusion: Conversations with Protoclassical Cinema

The historical examples I have enumerated here only scratch the surface of the telegraph's relationship to early film. Archival material yet to be uncovered on the topic may force radical revisions of my hypotheses. But the three approaches to this relationship that I have begun to map seem the most promising entryways into the study of early cinema as an intermedia phenomenon. The importance of telegraph-cinema research, and of the study of cinema's relationships to electrical media in general, resides in the insights it can offer into two especially tough historical questions. First, what was the temporal experience of early cinema like, compared to the experience of simultaneity between message and receiver experienced by users of electrical media?[59] And second, what effects, if any, did the discursive resonance between electrical communication and cinematic communication have on the development of classical narrative, and specifically on the changing relationship between audiences and screen that resulted as cinema stopped delivering "news" and concentrated almost exclusively on delivering stories?

These questions relate to each other closely in that they both ask about early cinema's *presence* to its viewers, by which I mean both how "live" the images seemed and the degree to which films were frankly presentational and reflexive. The dominant logic of film studies today leans toward the following position on this historical conundrum: If early film courted an impression of astonishing electrical instantaneity and possibilities for communication, then classical narrative films curtailed that impression in the process of eliminating the most obvious vestiges of the cinema of attractions. Following this logic, the self-referential tendencies of telegraphy and other electrical media, particularly evident in demonstrations and in the news services' aggrandizing self-consciousness about the telegraph's prowess, would make the telegraph a bad model for thinking about the cinema's cultural place once the early "cinema of attractions" began to wane. The industry had discovered by 1908 that the road to greater profits and middle-class acceptance lay in story films, which focused attention on representation and away from the apparatus. William Uricchio has suggested that early film viewers would have made a strict distinction between electrical media liveness and cinematic liveness from the beginning, understanding the former as a temporal category (referring to simultaneity between an action or message and its distant reception) and the latter as strictly a category of representation (the semblance of life offered by "moving" pictures).[60] If this were the case, early cinema would appear to have contained the seed of its future as strictly a medium of representation in its very status as a recording mechanism rather than a point-to-point transmitter.

But, to return to the point I made in the introduction about past approaches to electrical media and film, we need to look more closely at the social, economic, and historical positionings of media before we assume anything about what their technological capabilities signified to their users. The telegraph still carried an aura of instantaneity at the turn of the century, but that aura was tempered by the fact that only a tiny proportion of Americans enjoyed direct access to it. This uneven mixture of temporal and representational discourses—what ultimately mattered about telegraphic missives were the news stories or personal (though hardly private) messages they related, but their urgency would nonetheless have been greatly diminished without the discourse of instantaneity—makes the telegraph a kind of fraternal twin to the early cinema, especially if we consider the latter's uncanny relationship to viewers, its tendency to play with their understanding of where, and when, they sat in relation to the shocking and distracting screen images.

The most surprising thing that this parallel suggests, however, is that the illusionist form of "presence" conjured by classical film narration continued to draw on the same concept of cinematic communication as an overt and self-reflexive *act* that characterized the mode of address of living newspaper films and simulated "realities" like *The Life of an American Fireman*. In other words, the technological aspect of film and of its emerging classical mode of storytelling was not treated as an embarrassing reference to cinema's boisterous past—a scandal that the fig leaves of character and plot were engineered to obscure—but a focus of viewer interest and pleasure. I'd like to conclude by suggesting that we use telegraphic-cinematic discourse to help us rethink the transition to classical narrative form as a transition between variant definitions of *realism* that depend in part on telegraphy's own paradoxical position as both a spectacle in itself and a transmitter of distant messages or narratives: realism as a function of the spectator's confrontation with the photographic image (the "attractions" model); and realism as a function of cinematic "speech," an idea I will explain below.

The title given by the Edison company to the shot of actor Justus Barnes firing his pistol at the camera in *The Great Train Robbery*, "Realism," helps clarify how the cinema of attractions circa 1903 equated realism with the viscerality of film's effects, an equation also implied by the presentational and episodic nature of the rest of the film.[61] But in 1908, as the cinema of attractions was being replaced by one that emphasized dramatic illusions, an article in the trade periodical *Moving Picture World* asserted a much different definition of realism. The anonymous columnist recounted an "amusing incident in a New York theater" in which several spectators "involuntarily exclaimed, 'Don't drink that'" when a character seemed about to sip a poisoned drink (the film's title is not mentioned, but the suspenseful poisoning scene described here strongly suggests early Griffith). The reporter marveled:

> Surely manufacturers could not go farther than this in film realism. When they can induce those in their audience to warn characters not to do something they have accomplished what is most desirable. They have made the pictures speak. And the incident illustrates the close attention that is paid to a larger proportion of the films thrown on the screen. Even though they are mute the audience is as still as though the actors were actually speaking.[62]

The *Moving Picture World* reporter had a right to be excited about this development. One kind of cinematic "speech" that the journal had previously disparaged, that is, uncontrolled audience response, now seemed to be giving way to a different kind of conversation, one between screen and viewer that meshed neatly with the industry's interests because it privileged the screen as "speaker." Driven by pressure from reformers who complained of rowdy and unsavory exhibition spaces and by the economic imperative to gain greater control over exhibition for film producers, the Motion Picture Patents Company (headed by Edison and Biograph) began between 1907 and 1908 to seek alternatives to such extrafilmic narrational devices as live lecturers and even intertitles on the grounds that they distracted spectators from the stories unfolding on the screen.[63] But the reporter's comment does not entirely give up *discours* in the name of *histoire,* for while it casts narrative attentiveness and even absorption as desirable spectatorial habits, it lauds the speech-effect of the unnamed film specifically for the overtly interactive relationship it forges between screen and audience. As the reporter describes it, this new "realism" depended on the introduction of three kinds of speech into the viewing space: 1. the characters in the film speaking to each other; 2. the image speaking to the audience; and 3. the audience speaking back to the image as a sign of its involvement in the story. Though no longer speaking to other members of the crowd like the impromptu speechmaker at the McKinley film a decade earlier, the spectator (as viewed by the reporter) treats the film as a vehicle for information exchange rather than a representation pure and simple and asserts her own presence in the theater by responding with a message of her own. The film "talks" via images, and the ideal spectator proves her engagement by talking back.

The *Moving Picture World* reporter offers a surprising perspective on the emergence of classicality that suggests a slender thread of continuity between classical cinema and its distracting precursor, a lineage to which telegraphic discourse offers a possible key. Film images of telegraphs during the protoclassical era (1908–1917) rehearse both versions of realism I have described in that they continue to confront the audience with cinema's technological nature, while inviting narrative interest via the communications model the reporter suggests. Several examples in particular demonstrate the overlap between realist paradigms: the first chapter of the Mollie King serial *The Mystery of the Double Cross* (Pathé, 1917, d. William Parke), and the one- and two-reelers *The Telegraph Operators* (Éclair, 1915?), *The Dude Operator* (Edison, 1917, d. Saul Harrison), and *One Kind of Wireless* (Edison, 1917, d. Harrison).

The "Iron Claw" chapter of *Mystery of the Double Cross* contains a typical spectacle shot of a new kind of telegraph, the shipboard wireless set, in which the complicated-looking (authentic?) machinery takes up the left third of a medium shot of its operator. Though filmed in close-up, the telegraph's role is not substantially different from the one played by a similar set in Wallace McCutcheon's 1908 Biograph film *Caught by Wireless,* where the machinery grabs all the attention in the single tableau shot in which it appears. The last three examples, however, are more unusual. Although they promise telegraphy in their titles, the films themselves focus less on telegraphy than on the acts of encoding and decoding various kinds of messages. They do this by envisioning "code" in various ways: by following shots of the telegraph with intertitles whose letters appear one at a time (*The Dude Operator);* by showing trapped lovers, both skilled operators, tapping messages to each other on a window (*The Telegraph Operators);* and by semaphore and flashing lights, used by a "boy" operator to avert a railroad disaster (*One Kind of Wireless*). The effacement of the titular medium altogether in the latter two "telegraph" films might be an indicator of how closely the cinema identified itself with telegraphy even at this late date, when producers were increasingly underplaying the cinema as a technology and emphasizing story and character. By phasing out telegraphs, *The Telegraph Operators* and *One Kind of Wireless* metaphorically stress the "naturalness" of cinematic storytelling, while nevertheless remaining reliant on the *idea* of the telegraph for narrative interest, and thus extending the telegraphic subtext of early films into the mid-teens. The screen image, these movies tell us, is just as capable of sending specific, complex messages as is the telegraph, because this image also relies on encoding to get its point across. The impact of the dramas these films spin relies in part upon the dialectical relationship they claim between cinema and telegraph, as each film rehearses the older medium's insistence that witnessing the act of transmission is the ultimate guarantee of unbiased contact between sender—in this case the film producers and their narrative proxies, the characters—

and receiver. For the present, I will call the telegraphic tone of these films *communicative realism*—a form of reflexivity that defines cinematic realism in terms of the "live" process of decoding cinematic messages, that is, the active work of spectatorship.

Rather than think of the films I've just described as "stuck" somewhere between the cinema of attractions and classical cinema, I want to suggest that communicative realism is integral to a baseline definition of protoclassical cinema, whose mightiest proponents continually promoted cinema as a very self-conscious form of speech: the "universal language" of the photoplay. In 1917, the year that *One Kind of Wireless* was produced, early film theorist Vachel Lindsay delivered an address at Columbia University (reported by Epes W. Sargent in *Moving Picture World)* in which he amended the definition of film as a unique art form that he had first outlined in *The Art of the Moving Picture* (1915). Lindsay's speech reemphasized his book's argument that film should never attempt to appeal to the ear, like drama, but only to the eye.[64] At the same time, however, the concept of speech dominated Lindsay's

> new definition of the photoplay, [which he declared] to be *a conversation between two places,* using for his illustration the balcony scene from the Bushman-Bayne production of Romeo and Juliet in which the flashes alternate between Romeo and the balcony where Juliet is sitting. Perhaps the idea may be better suggested by saying that photoplay demands a story that is never held long in a single location as opposed to the limitation of the stage settings of the spoken drama.[65]

Like the 1908 *Moving Picture World* anecdote, Lindsay's address stresses representation over presentationality as the essence of cinematic communication, mobilizing the term "conversation" mainly as a metaphor for how editing connects characters within the film. At the same time, however, he casts the location changes caused by shot–reverse shot editing as the epitome of cinematic speech, not a device that effaces itself in the service of naturalism but a technical feat to be appreciated as such (in a similar vein, Griffith's editing had been touted by reviewers of *The Birth of a Nation* two years before as proof that the cinema was becoming a unique form of art).

Lindsay's use of interspatial conversation as the leading metaphor for his definition of film reminds us that in 1917 the cinema was still presenting *itself* and its powers over space and time nearly as much as it presented stories, and not without aspects of its electrical media legacy in tow. By describing the smooth continuity of narrative editing as a "conversation" taking place between "flashes" (shots), Linsday and Sargent summon the intermedia specters of both the telephone (point-to-point conversation) and the telegraph (with its transfer of messages in a flash, over the lightning lines) to help identify the new-found powers of this apparently noninteractive, photographic medium.[66] The telegraphic atmosphere surrounding films like *One Kind of Wireless* might be more important to Lindsay's definition of cinema than the telephone, however, because unlike the telephone, which delivered speech directly, telegraph and cinema transmitted messages in stages. Morse code had to be translated by an expert, and most dialogue in films had to be visualized in the form of intertitles, or inferred by the viewer who knew how to read visual codes such as pantomime, crosscutting, the close-up, and other devices which meant something quite different in their classical context than they had only a few years earlier. The "universal language" of cinema championed by Griffith, Lindsay, and many others in the 1910s—the new hieroglyphics, a picture language that would be instantly grasped by all—was nevertheless understood as a code that had to be self-consciously invented, taught, and learned before it could become naturalized.[67]

The focus on decoding in *One Kind of Wireless* and *The Telegraph Operators* visualizes Lindsay's theory of film as conversation by creating a parallel between the visual messages sent *in* the film and the visual messages sent *by* the film. Sargent's summary of Lindsay's speech leaves it a little ambiguous which two "places" Lindsay sees "conversing" in *Romeo and Juliet:* the places shown in shots A and B, or the imaginary space of the photoplay and the theater space. What makes the later telegraph films transitional in the history of the story film is that they remain overtly interested in the process of their own articulation (and invite the spectator's interest as well), while still honoring the industry's desire for representational transparency by motivating crosscuts in a more or less unobtrusive fashion. By 1917, industry and critics defined cinematic "realism" as the reduction of self-conscious spectacle, not its blatant recognition. And yet the acts of communication represented in films like *The Telegraph Operators* seem to be motivated by an aesthetic of telegraphic display more than a classical ideal of invisible storytelling. The intertitles carrying the literal message are utterly redundant and unnecessary to the narration; we cannot help but infer the message from the context, but the decoded message appears in the titles anyway. In the case of the two Edison films, the letters are revealed one at a time like individual Morse code signals unfurling in time before the well-trained operator. By displaying coded messages and translating them at the climactic moments, these scenes construct a makeshift analog to the self-referential authority of telegraphed news, supplementing the dramatic urgency with the familiar thrill of getting a news story hot from the wire.

The memory of the telegraph as a spectacle of communication seems to have been near enough in time to make these moments seem much less anomalous in 1917 than they do now, as we compare these films to the majority of mid-teens films, which appear to prefer unselfconscious story to self-conscious discourse. To conjure up the telegraph as these films of the 1910s do was to allegorize film's newly refined language of visual storytelling and turn it, paradoxically, into an attraction. If I am correct about the relative integrity of communicative realism as a moment in the development of classical cinema, then we are not so far from the operational aesthetic by 1917 after all, only as far as the public consumption of cinema was from the public consumption of the telegraph, and that distance was much shorter than we might have imagined. These films present their technological base and their textuality as factors that can peacefully coexist in the production of narrative entertainment. Indeed, the pleasure of watching early classical cinema may have depended in part on the feeling of being in on the act of constructing a new medium, learning a new technological language that had to be decoded by mobilizing not one but several kinds of inter- and extratextual knowledge. Viewers who wanted to involve themselves in the lives and stories of their new screen idols still had to decode what the pictures transmitted from afar were trying to say, and that act of decoding was a kind of labor that only "experts" in the language of cinematic narrative, not to mention that of the emerging star system with its dependence on extratextual publicity and scandals, could perform.

By inviting spectators to complete the circuit of filmic meaning by using their expertise in decoding sequences of images, the cinema rehearsed the old telegraphic expectations of universal access to media, and linked spectatorship with a discourse of free public exchange that continued to make the telegraph a symbol of (the possibility of) national unity even this late in its history. Whereas the privately owned telegraph business deflated fantasies of unity by blocking universal access, the equally privatized but literally collectivizing cinema found its niche in the kinds of mass "demonstrations" that had initially been crucial to the telegraph's cultural meaning but which became secondary to its more instrumental social functions. Without romanticizing the counter-public possibilities of film's second and third decades, I think it can be claimed that the cinema of the 1910s was still cast as a space of technologically sponsored exchange, in which the concept of cinematic "speech" and its interpretation hinged on the possibility of viewers' active, vocal engagement with films and fellow spectators, although the terms of that engagement were becoming more restrictive.[68]

Indeed, an important clue to the historical availability of the term *conversation* to describe shot-reverse shot editing may lie in the conceptual territory somewhere between electrical communication as an allegory for editing (and vice-versa), and Lindsay's fervent request in *The Art of the Moving Picture* that spectators remain vocal participants in the cinematic experience, as if they were (shades of Bertolt Brecht!) watching a sports event:

> At the [theater] door let each person be handed the following card:—
>
> "You are encouraged to discuss the picture with the friend who accompanies you to this place. Conversation, of course, must be sufficiently subdued not to disturb the stranger who did not come with you to the theater. If you are so disposed, consider your answers to these questions: What [photo]play or part of a play given in this theatre did you like the most today? . . ."
>
> . . . The fan at the photoplay, as at the base-ball grounds, is neither a low-brow nor a high-brow. . . . In both places he has the privilege of comment while the game goes on.[69]

Lindsay tries to distinguish his scenario from the all-too-public atmosphere of the cinema of attractions, urging the reader to keep "strangers" strange and to conceive of the viewing public as a group only in the empirical sense. But his description of cinema-literate fans "roasting the umpire" (the exhibitor or producer, as the spectator wills) retains the distracted relationship to films that characterized the attractions mode, and, like Benjamin in his own extrapolation of Brecht's spectator-as-sports-fan ideal, casts the viewer as a technological expert possessing a distanced critical eye.[70]

In this sense, at least, spectators in 1917 could yet have experienced films as demonstrations of a communications medium under development, its aesthetic, social, and political futures still undecided and certainly still intertwined. Addressed as technological experts simply because they were watching films—an unusual occurrence at a time when technological consumption increasingly stressed the importance of functional knowledge over a deep understanding of how things worked, and left expertise in the hands of the engineers—early spectators sat in the presence of a visual "telegraph" that offered them an unusual opportunity to get in on the conversations of technological modernity.

Acknowledgments

This essay was written with the help of a grant from the Georgia Tech Foundation. I would like to thank Joe Balian, Zoran Sinobad, Madeleine Matz, and especially Rosemary Hanes at the Library of Congress's Motion Picture, Broadcasting, and Recorded Sound Division for their aid during my visit in June 2000; the films I saw there were both electric and astonishing. Thanks also go to colleagues and friends who read and commented on drafts with great care: Richard Grusin, Jami Moss, Patrick Sharp, Lisa Yaszek, and Kate McNeal Young. Special gratitude is reserved for Lisa Gitelman and Geoff Pingree, whose comments and suggestions have been immeasurably helpful.

Notes

1. Gunning, *"The Lonely Villa," The Griffith Project,* vol. 2, ed. Paolo Cherchi Usai (London: BFI/ Le Giornate del Cinema Muto, 1999), 143. See also Gunning, "Heard over the Phone: *The Lonely Villa* and the de Lorde Tradition of the Terrors of Technology," *Screen* 32 (1991): 186–187, and Eileen Bowser, *The Transformation of Cinema: 1907–1915* (Berkeley: University of California Press, 1990), 64–71.

2. Raymond Bellour, "To Alternate/To Narrate," trans. Inge Pruks, *Early Cinema: Space/Frame/ Narrative,* ed. Thomas Elsaesser with Adam Barker (London: British Film Institute, 1990), 360–374.

3. Gunning, "Heard over the Phone," 186, 188–192, 193.

4. For the best sampling of theoretical and historical work on cinema's place in nineteenth-century visual culture, see *Cinema and the Invention of Modern Life,* ed. Leo Charney and Vanessa R. Schwartz (Berkeley: University of California Press, 1995).

5. This essay concludes in 1917 in order to coincide with the year cited by David Bordwell, Janet Staiger, and Kristin Thompson as the year the "classical Hollywood cinema" began to stabilize, as a set of narrative-stylistic conventions and as a method of industrial production that achieved efficiency and intelligibility through these conventions. Bordwell et al., *The Classical Hollywood Cinema: Film Style and Mode of Production to 1960* (New York: Columbia University Press, 1985).

6. If "the nation" is always both imagined and limited, as Benedict Anderson argues, then media like the telegraph and the cinema invite broader geographical participation in that shared fantasy while at the same time limiting participation to those who comprehend the lingua franca of the medium. By 1909, the cinema was expected to educate immigrants and thus invest them with an American identity precisely because its visual appeal seemed to circumvent the limits of language; but as I will relate in the conclusion, a discourse of visual symbolism kept notions of encoding and deciphering—and equally the notion of a community limited by shared knowledge—alive and well in discussions of cinema and politics. On national identity and print media, see Benedict Anderson, *Imagined Communities: Reflections on the Origin and Spread of Nationalism,* rev. ed. (London: Verso, 1991), 5–7 and 37–46.

7. See for example Gunning, "The Cinema of Attractions: Early Film, Its Spectator and the Avant-Garde," *Early Cinema* 58; "An Aesthetic of Astonishment: Early Film and the (In)credulous Spectator," *Film Theory and Criticism,* 5th ed., ed. Leo Braudy and Marshall Cohen (New York: Oxford University Press, 1999), 818–832.

8. See Charles Musser, *The Emergence of Cinema: The American Screen to 1907* (Berkeley: University of California Press, 1990), chapter 1 for an essential history of self-conscious "screen practices" from preindustrial Europe to nineteenth-century America. Hereafter I will refer to the book as *EC.*

9. Neil Harris, *Humbug: The Art of P. T. Barnum* (Boston: Little, Brown, 1973).

10. Charles Musser in collaboration with Carol Nelson, *High-Class Moving Pictures: Lyman H. Howe and the Forgotten Era of Traveling Exhibition, 1880–1920* (Princeton: Princeton University Press, 1991), 15, 56, 51.

11. Carolyn Marvin, *When Old Technologies Were New: Thinking About Electric Communication in the Late Nineteenth Century* (New York: Oxford University Press, 1988), 211.

12. David E. Nye, *American Technological Sublime* (Cambridge: MIT Press, 1994), 62.

13. When Leo Marx first introduces the "technological sublime," he invests the concept with political as well as epistemological ambiguity: "Back of the stock epithets and the pious, oracular tone" used by early American writers to describe new technologies and their effects on an uncultured landscape, "there are emotions which cannot be dismissed as mere hokum: a plausible incredulity, wonder, elation, and pride; a generous, humane delight at the promise of so much energy so soon to be released. But this is not to deny the intellectual hollowness of the rhetoric. The stock response to the panorama of progress, as Mill observed, by-passes ideas; it is essentially a buoyant feeling generated without words or thought." Marx, *The Machine in the Garden: Technology and the Pastoral Ideal in America* (New York: Oxford University Press, 1964), 206–207. The abstractness and ambivalence of the technological sublime return in the form of representation in early films with telegraphic pretensions and themes. As I will suggest later, such ambiguous representations gave audiences an opportunity to draw their own conclusions about the significance of telegraphy and cinema for visions of a "unified" nation.

14. K. G. Beauchamp, *Exhibiting Electricity* (London: Institution of Electrical Engineers, 1997), 1.

15. See Beauchamp, 92, 120, 127, 178, 191, 207–208.

16. Musser and Nelson, 56.

17. For an extensive discussion of the intermedial positioning of sound recording and film, see James Lastra, *Sound Technology and the American Cinema: Perception, Representation, Modernity* (New York: Columbia University Press, 2000). It seems appropriate that one of the earliest Edison Kinetoscope arcade film loop viewers installed in France could be found at the telegraph office of the *Petit Parisien,* 20 boulevard Montmarte. See Deac Russell, "'The new thing with the long name and the old thing with the name that isn't much shorter': A Chronology of Cinema 1889–1896," *Film History* 7 (1995): 128.

18. Gunning, "An Aesthetic," 819–824.

19. Walter Benjamin, "Paris, Capital of the Nineteenth Century," *Reflections,* ed. Peter Demetz, trans. Edmund Jephcott (New York, 1978), 152.

20. Gunning, "Phantom Images and Modern Manifestations: Spirit Photography, Magic Theater, Trick Films and Photography's Uncanny," *Fugitive Images: From Photography to Video,* ed. Patrice Petro (Bloomington: University of Indiana Press, 1995), 45, 66.

21. Richard J. Noakes, "Telegraphy is an occult art: Cromwell Fleetwood Varley and the diffusion of electricity to the other world," *British Journal for the History of Science* 32 (1999): 423, 425.

For an extremely useful discussion of telegraphic occultism in the US, see Jeffrey Sconce, *Haunted Media: Electronic Presence from Telegraphy to Television* (Durham: Duke University Press, 2000), chapter 1.

22. Noakes, 422; see also Sconce, 21.

23. Sconce, 27.

24. See Leo Marx's discussion of Webster's railroad speeches of 1847, in which the orator extols the steam engine and the new railway lines for their democratic force: They allow national access to goods previously restricted to one region; they close gaps of time and distance among the regions; and they are claimed (symbolically) by all Americans as "theirs." Marx, *The Machine in the Garden,* 209–214.

25. Alan J. Marcus and Howard P. Segal, *Technology in America: A Brief History* (San Diego: Harcourt Brace Jovanovich, 1989), 100.

26. George B. Prescott, *History Theory, and Practice of the Electric Telegraph,* 4th ed. (Boston, 1866), 215.

27. See Menahem Blondheim, *News Over the Wires: The Telegraph and the Flow of Public Information in America, 1844–1897* (Cambridge: Harvard University Press, 1994), 4 and 208–209 (note 13); Joel A. Tarr (with Thomas Finholt and David Goodman), "The City and the Telegraph: Urban Telecommunications in the Pre-Telephone Era," *Journal of Urban History* 14, no. 1 (1987): 50.

28. Richard B. Du Boff, "The Telegraph in Nineteenth-Century America: Technology and Monopoly," *Comparative Studies in Society and History* 26 (1984): 581.

29. Marvin, 221, 213–214.

30. Samuel Clemens, "Mrs. McWilliams and the Lightning" (1880), *The Complete Stories of Mark Twain,* ed. Charles Neider (New York: Bantam, 1990), 158.

31. Musser, *EC* 315–316; see also Musser's 1982 documentary film *Before the Nickelodeon,* which includes *Kansas Saloon Smashers* and a number of other "visual newspaper" films.

32. See Charles Musser, *Before the Nickelodeon: Edwin S. Porter and the Edison Manufacturing Company* (Berkeley: University of California Press, 1991), 162–63 (hereafter referred to as *BN*). Allen is quoted in Musser, *BN,* 162; see Robert C. Allen, "Contra the Chaser Theory," *Film Before Griffith,* ed. John Fell (Berkeley: University of California Press, 1983), 110.

33. Marvin, 221.

34. Jonathan Auerbach, "McKinley at Home: How Early American Cinema Made News, *American Quarterly* 51 (1999): 808–809.

35. Marvin reports fantasies circulating in the American and British presses by 1889 that promised we would "see by telegraph" as soon as Edison got around to inventing such a device. Marvin, *When Old Technologies Were New,* 197, 260. See also Mark Twain's story "From the London Times,

1904," which casts (what else?) a wry glance at such fantasies by returning them to the social context from which they were nearly always divorced.

36. Musser, *EC,* 241.

37. Prescott, v; Blondheim, 4.

38. Musser, *BN,* 69.

39. Auerbach, 813.

40. Discussing the political dimension of McKinley and William Jennings Bryan films of 1896, Musser says of William Heise's film *Bryan Train Scene at Orange* (Edison, 1896) that it was not "resolutely anti-Democratic," but neither was it "necessarily meant to be pro-Bryan. Rather it could be exhibited ambiguously and elicit both cheers and catcalls, generating an informal opinion poll from an audience." Musser, *BN,* 69.

41. Miriam Hansen distinguishes between *audience* and *spectator* as historical categories of cinema viewership: The first refers to viewers as "member[s] of a particular social group," while the second refers to the more ahistorical, universalizing, and isolated conception of the viewer that became the ideal for "social uplift" critics of early cinema and members of the film industry alike beginning roughly in 1908–1909. The burgeoning narrative cinema began to shed its rowdy fairground trappings for good only when it abstracted the empirical, historical viewer into an omniscient subject position via the formal and stylistic paradigm of classical cinema. See Hansen, *Babel and Babylon: Spectatorship in American Silent Film* (Cambridge: Harvard University Press, 1991), 84.

42. *New York World,* November 16, 1896, and *New Haven News,* November 23, 1896; rpt. in Niver, *Biograph Bulletins 1896–1908,* ed. Bebe Bergsten (Los Angeles, 1971), 18.

43. Auerbach, 811.

44. See Hansen, *Babel and Babylon,* chapters 1–2. Given that Hansen argues that the combination of the ambiguity of "attractions" films and the locally determined conditions of their exhibition *offered a structural possibility* for response and debate, Auerbach's assertion that his more deterministic and one-sided argument is "dialectical" compared to Hansen's (and Gunning's) is puzzling. Hansen asserts an ambiguous opportunity for early cinema to become an alternative public sphere for political discussion (see chapter 3) and shows how the institution of classical cinema limited this possibility at the level of the text. Auerbach, on the other hand, closes out this historical dialectic, in which self-conscious publicness pulls against an experience of "privatized," absorbed viewing, in favor of a claim that the latter had already taken hold of viewers by 1896.

45. Blondheim, 174; Du Boff, 582, 583.

46. Blondheim, 193.

47. Marvin, 107. For information on Alexander Graham Bell's 1876 telephone concerts and the public accessibility of telephones by 1902 (more than 2 million phones in the United States, three

and a half percent of which were pay phones), see Sidney H. Aronson, "Bell's Electrical Toy: What's the Use? The Sociology of Early Telephone Usage," *The Social Impact of the Telephone,* ed. Ithiel de Sola Pool (Cambridge: MIT Press, 1977), 20, 32.

48. See Noël Burch, "Porter, or Ambivalence," *Screen* 19, no. 4 (1978–1979): 91–105.

49. On the saloon as social hub of "workingmen's leisure," see Kathy Peiss, *Cheap Amusements: Working Women and Leisure in Turn-of-the-Century New York* (Philadelphia: Temple University Press, 1986), 16–18.

50. Blondheim, 36.

51. André Gaudreault, "Detours in Film Narrative: The Development of Cross-Cutting," trans. Charles Musser and Martin Sopocy, *Early Cinema,* 139.

52. Tarr et al., "The City and the Telegraph," 59.

53. Firefighting was a popular topic in many forms of turn-of-the-century mass culture, regularly sensationalized and sentimentalized in fiction, poetry, newspapers, popular engravings, plays, magic lantern shows, and films. See Musser, *BN,* 218–221. On the showcasing of various re-hearsed disasters at Coney and elsewhere, and their function as a stimulus shield for the subjects of modernity, see John F. Kasson, *Amusing the Million: Coney Island at the Turn of the Century* (New York: Hill and Wang, 1978), 71–72.

54. See Musser, *BN,* 216.

55. "The film [*The Life of an American Fireman*] opens with a montage often comprising ellipses so startling that one wonders how audiences could follow the story without a lecturer's help" (Burch, 103).

56. Tarr, 53, 59.

57. The scene is infamous because the print acquired by the Museum of Modern Art in 1944 dif-fers dramatically from the Library of Congress copyright print. In the former, the climactic scene is crosscut so that narrative time unfolds continuously between the exterior and interior of the house. André Gaudreault convincingly argues that the crosscut version was reedited much later, and that the copyright version is probably close to the version distributed in 1903. See Gaud-reault, "Detours in Film Narrative,"133, 146–147, and passim. The copyright print's version of the final event fits more nearly with the structure of the film as a whole, which privileges seriality over narrative causality.

58. Musser, *BN,* 222.

59. I am referring to the temporal relationships early audiences perceived between the profilmic event and the film of that event. For an examination of chronology and temporality as they oper-ate *within* early narratives, see Gunning, "'Now You See It, Now You Don't': The Temporality of the Cinema of Attractions," *Silent Film,* ed. Richard Abel (New Brunswick: Rutgers University Press, 1996), 71–84.

60. Uricchio, "Television, Film and the Struggle for Media Identity," *Film History* 10 (1998): 119, 125.

61. Charles Musser argues that the "Realism" label placed on this shot indicates the relevance of "identification and emotional involvement with the drama" to the discourse of filmic realism even as early as 1903. My position obviously differs, but I am also suggesting that no single definition of realism overshadowed all the others at the time. Instead, "realism" appears to have been as heterogeneous and contradictory a category in early cinema as it was historically contingent. See Musser, "The Travel Genre in 1903–1914; Moving toward Fictional Narrative," *Early Cinema,* 130.

62. "Film Realism," *Moving Picture World* 3, no. 22 (November 28, 1908): 427.

63. For the key discussion of the industry's attempts to stabilize itself through production practices in 1908–1909, see Gunning, "Weaving a Narrative: Style and Economic Background in Griffith's 1908–1909 Biograph Films," *Early Cinema,* 336–340.

The use of intertitles in silent film has sometimes led scholars and students to assume that early cinema tagged its own silence as lack, acknowledging sound and especially speech as elements without which film was incomplete. In fact, however, the intertitle was considered a necessary evil—"evil" because it drew too much attention to itself. Eileen Bowser helps to debunk the myth of lack by outlining a debate that began around 1907 in the trade press about the relative value of intertitling to storytelling. The title allowed companies to confer complex story information that, at the time, could not be expressed clearly through any other means (editing being the most important possibility, one that Griffith would explore in new directions beginning the following year), and so served the industry's interests in cloaking its products in respectability and sophistication, and gaining greater control over the meanings of pictures. But the intertitle clashed with the industry and trade press's growing concern to make pictures "artistic" in their own right and, perhaps most surprisingly, its status as a visual interruption was thought to break the increasingly important illusion of reality (Bowser, *The Transformation of Cinema,* 140).

64. Almost nothing annoyed Lindsay more about film in 1915 than the use of stray music, unsanctioned by the filmmakers, to accompany screenings. See Linsday, *The Art of the Moving Picture* (1915, rev. ed. 1922; New York: Liveright, 1970), chapter 14.

65. Epes W. Sargent, "Vachel Lindsay on the Photoplay," *Moving Picture World* 31, no. 10 (March 10, 1917): 1583 (my emphasis).

66. Although it may seem something of a stretch to align "flashes" with "lightning lines," early names for the film camera appear to be etymologically linked to idiomatic names for the telegraph. Musser reprints two separate stories from the *Newark Daily News* detailing the shooting of *Life of an American Fireman* that use peculiarly telegraphic words to name the kinetograph: "chain-lightning cameras" and "lightning cameras." Musser, *BN,* 212 and 213.

67. A good example of the natural/learned paradox lodged in the concept of film as universal language can be found in a 1910 *Moving Picture World* story by the Reverend William Henry Jackson, cited by Miriam Hansen: "The ear may comprehend but one language, the eye comprehends all languages; [many] races may sit side by side and together read in the universal language of

the eye the selfsame subject," Jackson writes, implying that moving pictures may be automatically comprehended worldwide. But he acknowledges the need for conscious development when he avers that "we are still in the A B C of a new field; the alphabet must become a language, the toy will become a worldwide utility." Hansen, "Universal Language and Democratic Culture: Myths of Origin in Early American Cinema," *Myth and Enlightenment in American Literature,* ed. Dieter Meindl and Friedrich W. Horlacher with Martin Christadler (Erlangen: Universitätsbund Erlangen-Nürnberg 1985), 328.

68.   This claim implies the extension of Hansen's argument about the counter-public structure of early film exhibition beyond *The Birth of a Nation* and *Intolerance* (1916), films that each signaled the waning of the cinema of attractions in their own way (although as Hansen shows, the star system has kept the possibility of "excessive" viewing practices very much alive since then). For Hansen, the social and discursive heterogeneity of the nickelodeon enabled audiences, particularly immigrants, to recognize the perceptual and political fundamentals of their life experience: "The nickelodeon . . . opened up into a fantastic space, giving pleasure in the juxtaposition of diverse, often incompatible, and at times impossible sites or sights. . . . The aesthetics of disjunction not only contested the presumed homogeneity of dominant culture and society in the name of which immigrants were marginalized and alienated; more important, it lent the experience of disorientation and displacement the objectivity of collective expression." Hansen, *Babel and Babylon,* 108.

69.   Lindsay, 225, 227.

70.   "The film makes the cult value [of the work of art] recede into the background not only by putting the public in the position of the critic, but also by the fact that at the movies this position requires no attention. The public is an examiner, but an absent-minded one." Benjamin, "The Work of Art in the Age of Mechanical Reproduction," *Illuminations,* ed. Hannah Arendt, trans. Harry Zohn (New York: Schocken, 1968), 240–241.

Contributors

Wendy Bellion is assistant professor of art history at Rutgers, the State University of New Jersey. She holds a Ph.D. from Northwestern University and is writing a book about visual deception and political culture in the early American republic.

Erin C. Blake completed her Ph.D. in art history at Stanford University. She is the Curator of Art at the Folger Shakespeare Library in Washington.

Patricia Crain is associate professor of English at the University of Minnesota. She is the author of *The Story of A: Alphabetization in America from* The New England Primer *to* The Scarlet Letter (Stanford, 2000).

Ellen Gruber Garvey is associate professor of English at New Jersey City University. She is the author of *The Adman in the Parlor: Magazines and the Gendering of Consumer Culture, 1880s to 1910s* (Oxford, 1996), which won the Best Book prize for 1996 from the Society for the History of Authorship, Reading, and Publishing.

Lisa Gitelman is assistant professor of English and media studies at Catholic University. She is the author of *Scripts, Grooves and Writing Machines: Representing Technology in the Edison Era* (Stanford, 1999).

Geoffrey B. Pingree is assistant professor of cinema studies and English at Oberlin College. He has written on documentary film and on Spanish cinema, and he is completing a book entitled *Documentary Film and National Identity in Spain*.

Gregory Radick lectures in history and philosophy of science at the University of Leeds. He is coeditor (with Jonathan Hodge) of the forthcoming *Cambridge Companion to Darwin*. He is currently working on a study of R. L. Garner's primate "playback" experiments, their disappearance from the scientific repertoire in the early twentieth century, and their subsequent reinvention and development.

Laura Burd Schiavo, a doctoral candidate in American Studies at George Washington University, is on the staff of the City Museum of The Historical Society of Washington, D.C. She has held research fellowships at the Library Company of Philadelphia, the American Antiquarian Society, and the Smithsonian Institution and is completing a dissertation entitled "'A Collection of Endless Extent and Beauty': Stereographs, Perception, Taste and the American Middle Class, 1850–1882."

Katherine Stubbs is associate professor of English at Colby College. Her book, *Fantasies of Fluidity: Women and Class Passing in the American Text,* is forthcoming from Duke University Press.

Diane Zimmerman Umble is professor and chair of the Department of Communication and Theatre at Millersville University. She is the author of *Holding the Line: The Telephone in Old Order Mennonite and Amish Life* (Johns Hopkins, 1996).

Paul Young teaches film and media studies, mass culture, and literature in the English department at the University of Missouri-Columbia. His book, *The Cinema Dreams Its Rivals: New Media and Hollywood's Public Spheres,* is forthcoming from the University of Minnesota Press. He has also published articles on cyberspace in 1990s film and on writing the cultural history of film noir.

Index